POLYMER SCIENCE AND TECHNOLOGY
Volume 11

# POLYMER ALLOYS II

Blends, Blocks, Grafts, and
Interpenetrating Networks

# POLYMER SCIENCE AND TECHNOLOGY

A Continuation Order Plan is available for this series. A continuation order will bring
delivery of each new volume immediately upon publication. Volumes are billed only upon
actual shipment. For further information please contact the publisher.

**POLYMER SCIENCE AND TECHNOLOGY**
Volume 11

# POLYMER ALLOYS II
Blends, Blocks, Grafts, and
Interpenetrating Networks

Edited by

## Daniel Klempner
## and Kurt C. Frisch

*Polymer Institute*
*University of Detroit*

SPRINGER SCIENCE+BUSINESS MEDIA, LLC

Library of Congress Cataloging in Publication Data

Symposium on Polymer Alloys, Honolulu, 1979.
  Polymer alloys II.

  (Polymer science and technology; v. 11)
  Includes index.
  1. Polymers and polymerization—Congresses. I. Klempner, Daniel. II. Frisch, Kurt
Charles, 1918-      III. Title. IV. Series.
QD380.S939  1979                          547.7                          79-28487
  ISBN 978-1-4684-3631-0          ISBN 978-1-4684-3629-7 (eBook)
  DOI 10.1007/978-1-4684-3629-7

Proceedings of the Symposium on Polymer Alloys, held in
Honolulu, Hawaii, April, 1979.

© Springer Science+Business Media New York 1980
Originally published by Plenum Press, New York in 1980
Softcover reprint of the hardcover 1st edition 1980

# Preface

The term "alloy" as pertaining to polymers has become an increasingly popular description of composites of polymers, particularly since the publication of the first volume in this series in 1977. Polymer alloy refers to that class of macromolecular materials which, in general, consists of combinations of chemically different polymers.

The polymers involved in these combinations may be heterogeneous (multiphase) or homogeneous (single phase). They may be linked together with covalent bonds between the component polymers (block copolymers, graft copolymers), linked topologically with no covalent bonds (interpenetrating polymer networks), or not linked at all except physically (polyblends). In addition, they may be linear (thermoplastic), crosslinked (thermosetting), crystalline, or amorphous, although the latter is more common.

To the immense satisfaction - but not surprise - of the editors, there has been no decrease in the research and development of polymer alloys since the publication of the first volume, as evidenced by numerous publications, conferences and symposia. Continued advances in polymer technology caused by the design of new types of polymer alloys have also been noted. This technological interest stems from the fact that these materials very often exhibit a synergism in properties achievable only by the formation of polymer alloys. The classic examples, of course, are the high impact plastics, which are either polyblends, block, or graft copolymers composed of a rubbery and a glassy polymer. Interpenetrating polymer networks (IPN's) of such polymers also exhibit the same, or even greater, synergism.

This book presents the proceedings of the Symposium on Polymer Alloys, sponsored by the American Chemical Society's Division of Organic Coatings and Plastics Chemistry, held at the American Chemical Society/Chemical Society of Japan Chemical Congress in Honolulu, Hawaii, in April, 1979. It is representative of the most recent efforts of both American and foreign scientists in a field of increasing interest:  the preparation, character- ization and properties of polymer alloys.

The editors wish to take this opportunity to express their gratitude to the authors who contributed to this book and to the University of Detroit for its encouragement of this effort.

Daniel Klempner

Kurt C. Frisch

Polymer Institute

University of Detroit

# Contents

SECTION IV.   POLYBLENDS

MORPHOLOGY AND PHASE RELATIONSHIPS OF LOW-MOLECULAR-WEIGHT
POLYSTYRENE IN POLY(METHYL METHACRYLATE) AND METHYL METHACRYLATE/
STYRENE COPOLYMERS

Edward V. Thompson

Department of Chemical Engineering
University of Maine at Orono, Orono, Maine   04473

INTRODUCTION

Over the past twenty years or more there has been widespread
interest in various kinds of multicomponent polymer systems, including polyblends, block copolymers, and segmented elastomers.
More recently, considerable interest has also been focused on
questions concerning polymer/polymer compatibility and incompatibility in these systems, and also in related systems such as interpenetrating networks and alloys.  Among this diverse group of multicomponent and in some cases multiphase polymers, one that has received perhaps the most overall attention, because of the combination of its great commercial importance and scientific interest, has
been the so-called high-impact plastics and resins.  Typically, these
multiphase polyblends are based on a dispersed, rubbery phase such
as polybutadiene which is contained in a glassy, continuous matrix
such as polystyrene.  The presence of the second, dispersed phase
imparts added impact strength, as measured, for example, by an Izod
impact apparatus, to the composite above that possessed by the homopolymer polystyrene itself; and it is this enhancement, of course,
which leads to the commercial importance of polyblends of this type.
In the development of high-impact plastics and the subsequent study
of their physical and mechanical properties, a considerable body of
information and data has appeared.  In particular, it is well known
that the size of the dispersed phase, as well as, among other variables, molecular weight and overall composition in the composite,
has a profound effect on the ultimate impact strength.  However, in
spite of this interest, very little systematic data have appeared,
especially concerning the effect of molecular weight and concentration of the dispersed phase on the phase relationships and final
properties of the system.  It is, therefore, the purpose of this and

1

several other reports to study the variation of molecular weight and concentration in the polystyrene/poly(methyl methacrylate) system to help elucidate these effects; and, in particular, their influence on morphology and phase relationships of the resulting multiphase system and on the size of the dispersed phase. Before proceeding, however, it seems appropriate to review briefly prior work by others which relates to the present discussion.

Multiphase polymer/polymer systems have been known since at least the turn of the century (1), although interest in questions concerning polymer/polymer compatibility, or incompatibility as is the more usual case, did not begin to develop until the late 1940's (2). This interest paralleled the commercialization of polymer systems based on two incompatible polymers, and more recently numerous technical publications and several volumes devoted to articles and reviews have appeared (3-8). We are especially interested here in morphological studies of multiphase polymer systems, and note that the first microscopic studies were probably made by Claver and Merz in connection with their studies of high-impact polystyrene (9). Traylor reported a method of investigating multiphase systems using phase contrast microscopy (10), and transmission electron microscopy has been successfully employed by several investigators (11-14).

It is well known that both the size of the particles forming the dispersed phase (15, 16) and the overall amount of the dispersed phase (17) are of crucial importance in determining the physical and mechanical properties of high-impact plastics, and, in particular, their impact strengths. These observations are generally thought to be related to the idea that the dispersed particles absorb and dissipate energy of impact by the formation of numerous, small internal crazes (18, 19). Several reports have appeared concerning the influence of molecular weight and composition of the dispersed phase on particle size. For example, Baer found that for molecular weights of polybutadiene above 60,000 the particle size increased linearly with molecular weight when plotted on a log-log chart (20). On the other hand, Moore showed that the concentration of the species which ultimately formed the dispersed phase was an important factor in predicting the particle size (21). These data are, however, incomplete since in the former case only variation of molecular weight was considered, and in the latter only variation of concentration. Further, the results do not lead to an overall predictive scheme for indicating whether an increase in molecular weight and/or composition will lead to increase or decrease in particle size. In fact, there appear to be contradictory situations where sometimes an increased particle size is noted, while in others a decreased size.

In a related area, questions involving compatibility and incompatibility of polymer/polymer and polymer/polymer/solvent systems have been investigated. Further, although perhaps to a more limited degree, reports have also appeared concerning the phase relation-

ships and binary and ternary phase diagrams of these systems.  In
an important series of reports, Molau and co-workers (22-27) investigated systems formed by first dissolving a polymeric species in a
monomer and then polymerizing the monomer to form a multicomponent
polymer system.  During polymerizations of this type an inversion
of phases was observed, and a phase inversion mechanism was proposed
to explain the phenomenon.  This phase inversion or a similar process has been suggested as the mechanism whereby polybutadiene becomes the dispersed phase in the production of high-impact plastics
based on butadiene and styrene (25, 28).  More recently, Kruse has
suggested that the phase inversion point may be interpreted in terms
of a ternary phase diagram for the polymer/polymer/monomer system
(29), and this is the point of view that we have advanced.  Finally,
it seems appropriate to mention the recent interest in polymer/poly-
mer compatibility (30), and in particular the investigations of
Massa who studied the polystyrene/poly(methyl methacrylate) system
(31).  This latter work will be discussed in a later section as it
relates to the present report.

    In the preceeding discussion we have noted the relative lack
of data correlating both the molecular weight and concentration of
the dispersed phase with particle size and ultimately mechanical
properties.  Because of this we have undertaken a study of the poly-
styrene/poly(methyl methacrylate) system to help to elucidate these
relationships.  We have previously reported on the low-molecular-
weight polystyrene in poly(methyl methacrylate) system (32, 33) and
the low-molecular-weight poly(methyl methacrylate) in polystyrene
system (34, 35).  More recently, we have discussed in a brief report
the low-molecular-weight polystyrene in styrene/methyl methacrylate
copolymer system (36).  It is the purpose of this report to review
our work concerning the low-molecular-weight polystyrene in poly-
(methyl methacrylate) system and, more importantly, to discuss our
new findings pertaining to the low-molecular-weight  polystyrene
in styrene/methyl methacrylate copolymer system.  The results of the
morphological study of the various multicomponent polymer specimens
using scanning electron microscopy are reported, and these results
discussed in terms of polystyrene/poly(methyl methacrylate-styrene)/
methyl methacrylate-styrene ternary phase diagrams.


EXPERIMENTAL PROCEDURES

    The investigations discussed in this report deal with the
polystyrene/poly(methyl methacrylate) system.  This pair of polymers
was selected for our initial studies for several reasons, the most
important being: (1) well-characterized samples of low-molecular-
weight, monodisperse polystyrene are readily available commercially;
and (2) the polystyrene/poly(methyl methacrylate) system is especially amenable to study by scanning electron microscope techniques,
as we will discuss in more detail below.  However, we anticipate

that other polymer/polymer systems can equally well be fabricated
and studied by techniques similar to those to be described below.
In fact, we are presently working on the technologically important
polybutadiene/polystyrene system, and expect to report our findings
in the near future.

Styrene (hereafter denoted S) (Eastman Kodak) and methyl methacrylate (hereafter denoted MMA) (Eastman Kodak) were dried and distilled under vacuum following standard procedures for purifying
monomers.  Polystyrene (hereafter denoted PS) samples of number-
average molecular weight ($M_n$) 2100 (Pressure Chemical, Pittsburg,
Pa.); 3100, 9600, and 19,650 (Waters Associates, Milford, Mass.);
and 3100, 12,000, 15,100, 18,900, and 50,400 (ArRo Laboratories,
Joliet, Ill.) were used as received.  The weight-average to number-
average molecular weights of these samples ranged from 1.02 to 1.10,
and thus can be considered to be reasonably monodisperse.  The initiator used for the polymerizations, 2,2'-azobis(isobutyronitrile)
(AIBN) (Eastman Kodak), was recrystallized from toluene.

Samples were prepared by first weighing a given amount of low-
molecular-weight PS into a 5-mm-i.d. Pyrex tube sealed at one end.
To this was then added with a long-needled hypodermic syringe the
appropriate amount of monomer to give the desired final composition.
The monomer consisted of either pure MMA or one of three MMA-S
mixtures (actual compositions: 74.99 wt % MMA and 25.01 wt % S;
50.00 wt % MMA and 50.00 wt % S; and 25.00 wt % MMA and 75.00 wt %
S) to which had been added 1/4 wt % AIBN.  After monomer addition
the tube was reweighed, and the open end then sealed.  Solution of
the PS in the MMA-S mixture was accomplished by physical agitation
of the tube, this process taking anywhere from a few minutes to
several days depending on the molecular weight and weight percent
of PS.  After complete solution had taken place, the monomer was
polymerized at 50°C for 2 days, followed by postcuring for about
12 hours at 80°C and 2 to 3 hours at 100°C.  Finally, the polymer/
polymer composite was recovered by shattering the glass tube.

The following points concerning the characterization of the
samples obtained by the procedure outlined above might be noted,
namely:
  (1) The samples are completely soluble in typical PS and PMMA
      solvents such as toluene or tetrahydrofuran.
  (2) In polymerizations of the type described above, no residual
      monomer can be detected by gas-liquid chromatographic
      techniques, and is assumed to be no greater than 0.15 wt %.
  (3) The molecular weight, as inferred from intrinsic viscosity
      measurements, of the PMMA or P(MMA-S) phase is very high,
      being 1,000,000 or more.
  (4) Comparisons of gel permeation chromatographs of actual samples
      and standards of identical composition indicated no evidence
      of grafting.

The morphological details and phase relationships of the final polymer/polymer systems were examined using a Stereoscan S4 (Cambridge Scientific Instruments, Ltd.) scanning electron microscope. The samples were brittle fractured, and mounted and examined following standard SEM techniques (typically, a 100-Å-thick gold coating to provide proper surface conduction; and microscope operation at 20-kV acceleration potential, 180 to 220-μA beam current, and 2.3 to 2.5-A filament current). Magnifications ranged from several hundred to over 10,000X.

The successful application of scanning electron microscopy to the study of morphology and phase relationships of two-phase PS/P(MMA-S) systems obviously presupposes techniques for distinquishing unambiguously the PS and P(MMA-S) regions. Pure PMMA, P(MMA-S) copolymers, and PS show characteristic fracture-surface features, and these serve as the basic criterion. For example, pure PMMA is characterized by rather regular parabolic structures with lines radiating from an area near the focus and pure PS by a much more irregular and in some cases distinctly three-dimensional appearance (37), while P(MMA-S) copolymers display surface features which are more-or-less intermediate. These morphologies are shown in Figures 1 through 5, where we note that P(75MMA-25S) and P(50MMA-50S) have morphologies very similar to PMMA whereas P(25MMA-75S) has a morphology essentially identical to PS. Further-

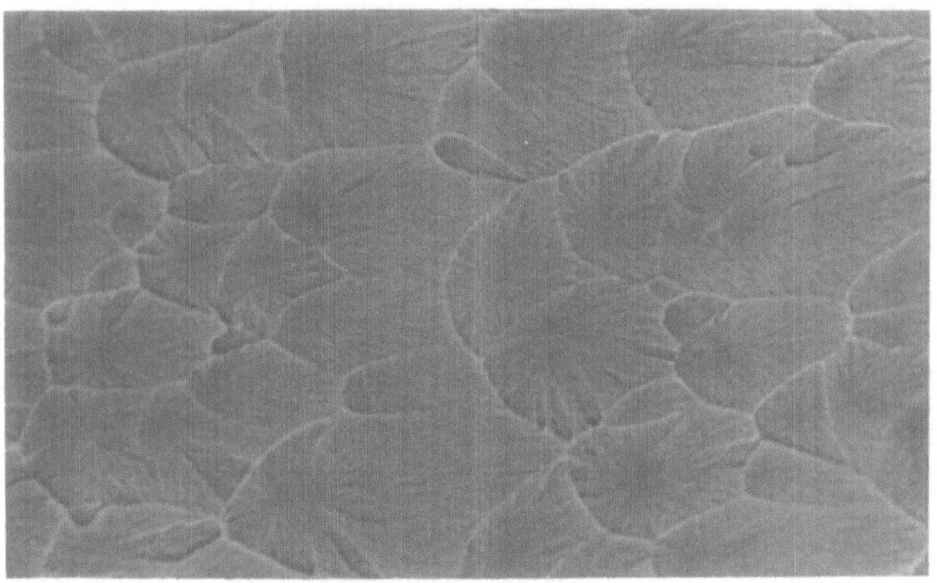

Fig. 1. Pure high-molecular-weight PMMA.   540X.

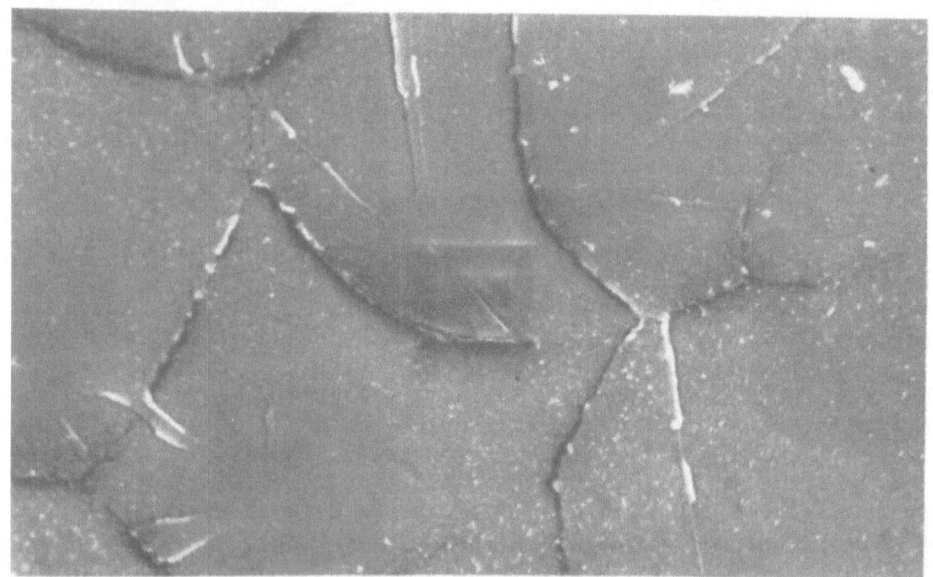

Fig. 2. Pure high-molecular-weight P(75MMA-25S).  540X.

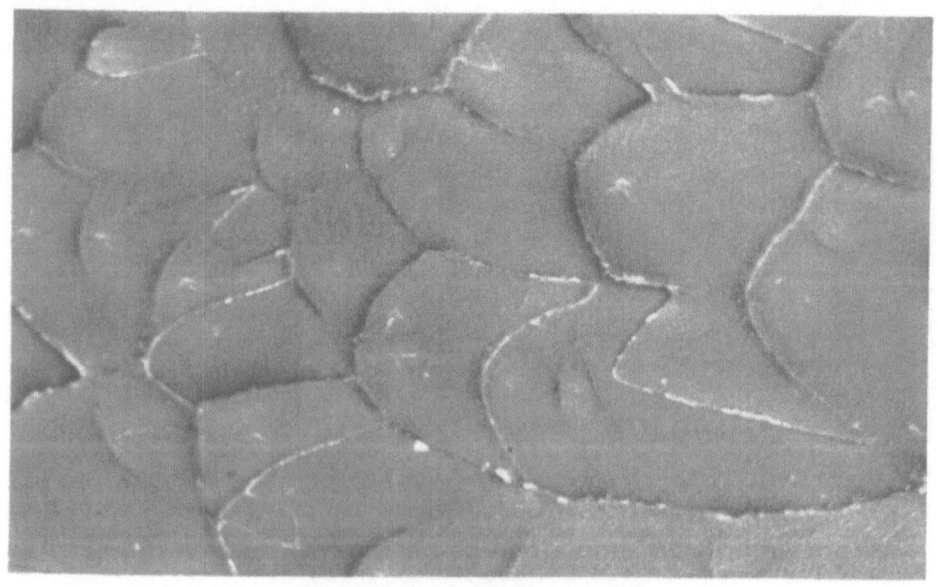

Fig. 3.  Pure high-molecular-weight P(50MMA-50S).  550X.

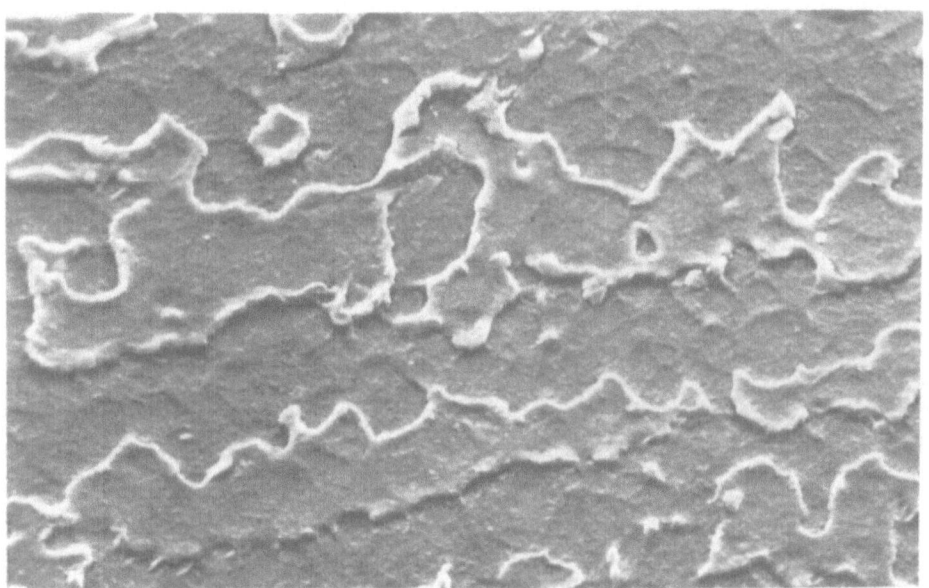

Fig. 4. Pure high-molecular-weight P(25MMA-75S).   2035X.

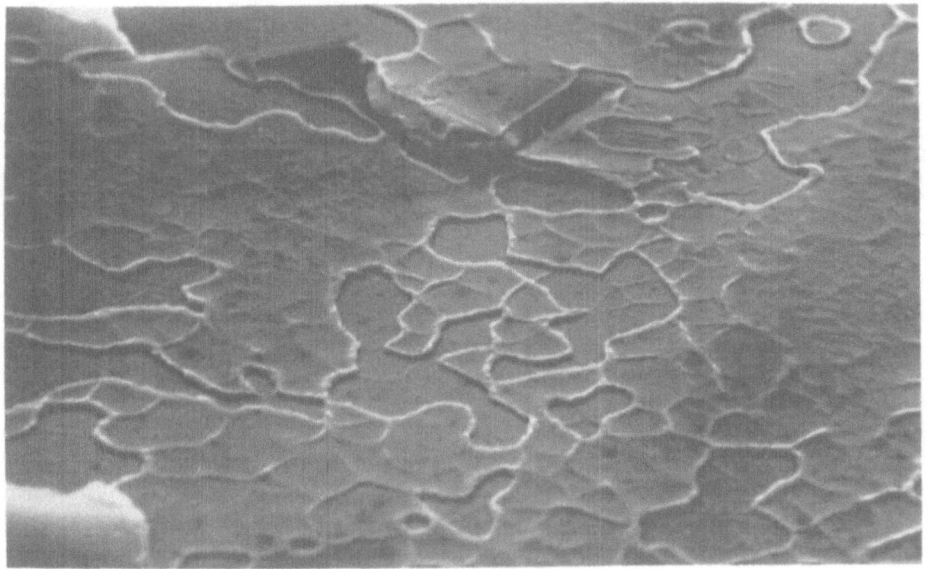

Fig. 5. Pure high-molecular-weight PS.   1150X.

more, these basic morphologies are retained when dispersed particles
are present, as shown in Figures 6 through 9, where in Figures 6
and 7 the continuous phases are PMMA and P(75MMA-25S) displaying the
morphologies characteristic of PMMA, while in Figures 8 and 9 the
continuous phases are PS and P(25MMA-75S) displaying the morphologies characteristic of PS.  Thus it is usually possible, based on
the overall surface morphology, to identify both the continuous and
dispersed phases unambiguously.  However, a second criterion is also
available for some samples.  It is well known that PMMA depolymer-
izes under the influence of an electron beam (38), and this is true
of PMMA samples which are exposed for prolonged periods of time to
the SEM electron beam, especially at magnifications of 5000X or
higher.  This phenomenon provides a very sensitive criterion for
identifying PMMA either as the continuous or dispersed phase, or
even as a subinclusion or multiple emulsion in a dispersed phase.
This criterion was used extensively in our earlier work involving
low-molecular-weight PS in PMMA and low-molecular-weight PMMA in PS,
and is discussed in considerable detail in reference 35.  It has
been possible to extend the experience gained in these earlier
studies to the present work in several cases where the surface morphologies themselves did not serve to unambiguously identify the
phases.

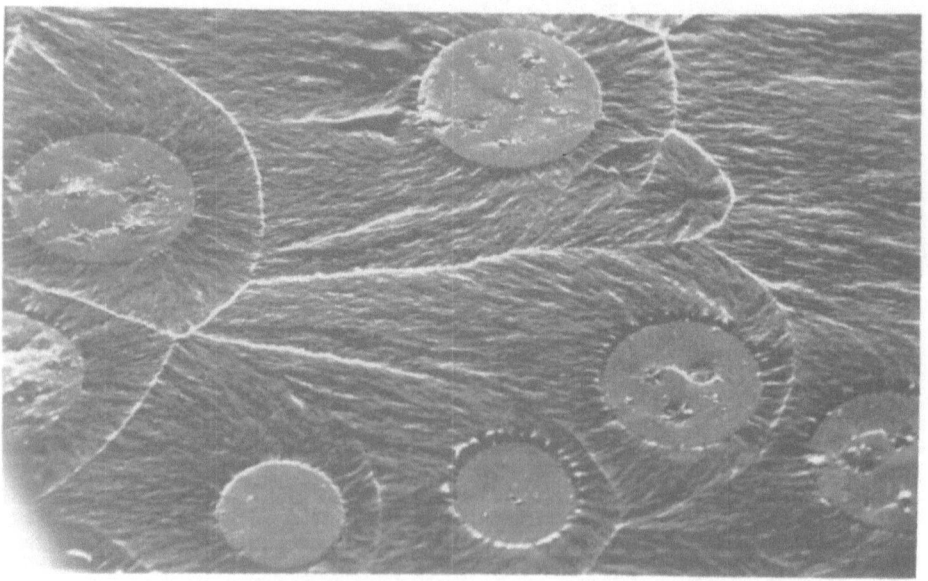

Fig. 6. PS/PMMA system with PS ($M_n$ = 19,650; W = 17.65 wt %) the
         dispersed phase and PMMA the continuous phase.  850X.

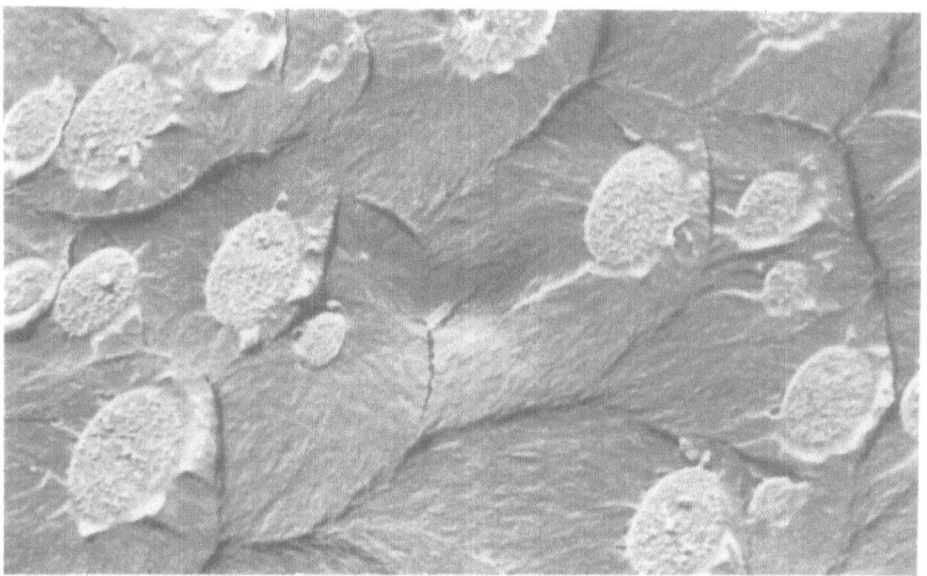

Fig. 7. PS/P(75MMA-25S) system with PS ($M_n$ = 50,400; W = 4.97 wt %) the dispersed phase and P(75MMA-25S) the continuous phase. 2015X.

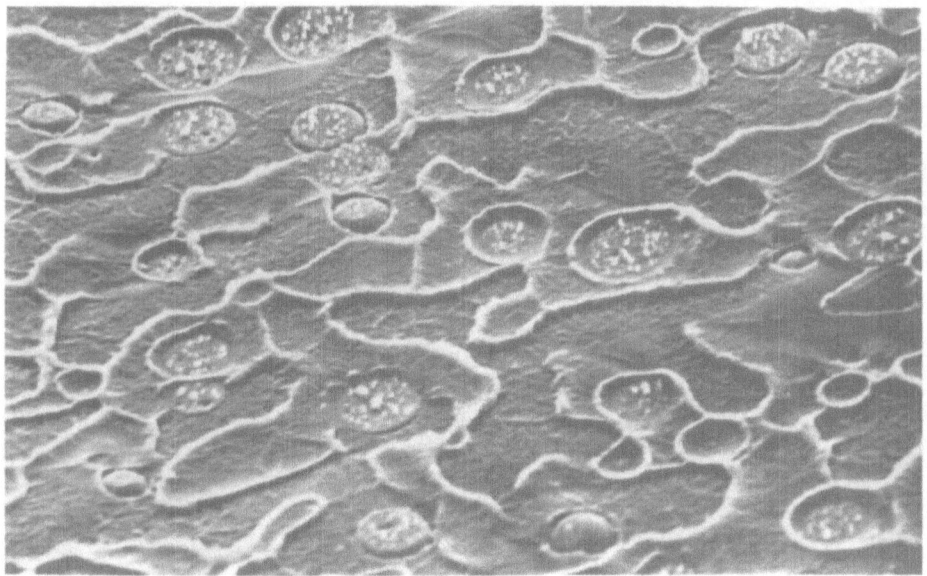

Fig. 8. PMMA/PS system with PMMA ($M_n$ = 5700; W = 20.13 wt %) the dispersed phase and PS the continuous phase.　1575X.

EXPERIMENTAL RESULTS

In Table I is listed a summary of the specimens prepared in conjunction with our earlier study of the low-molecular-weight PS in PMMA system; and in Table II the results of the topic of this report, the low-molecular-weight PS in P(MMA-S) system. Included is the number-average molecular weight ($M_n$) and weight percent (W) of the PS present in the initial PS/MMA or PS/MMA-S mixture, the comonomer composition (in Table II), and an indication of opacity, translucency, or transparency of the final specimen. It should be noted that many of the specimens listed in the PMMA column in Table II are replicas of those reported in Table I. We were interested in the reproducibility of our fabrication techniques, and an inspection of similar specimens shows virtually identical results with only two or three minor differences.

For the PS/PMMA and PS/P(MMA-S) systems under consideration, four distinct types of morphologies and phase relationships are observed, namely;

Type [1]. A single phase which displays the morphological features of pure PMMA or P(MMA-S) and in which no evidence of a dispersed phase can be detected at magnifications as high as 20,000X. Specimens which fall into this category produced SEM photomicrographs identical or

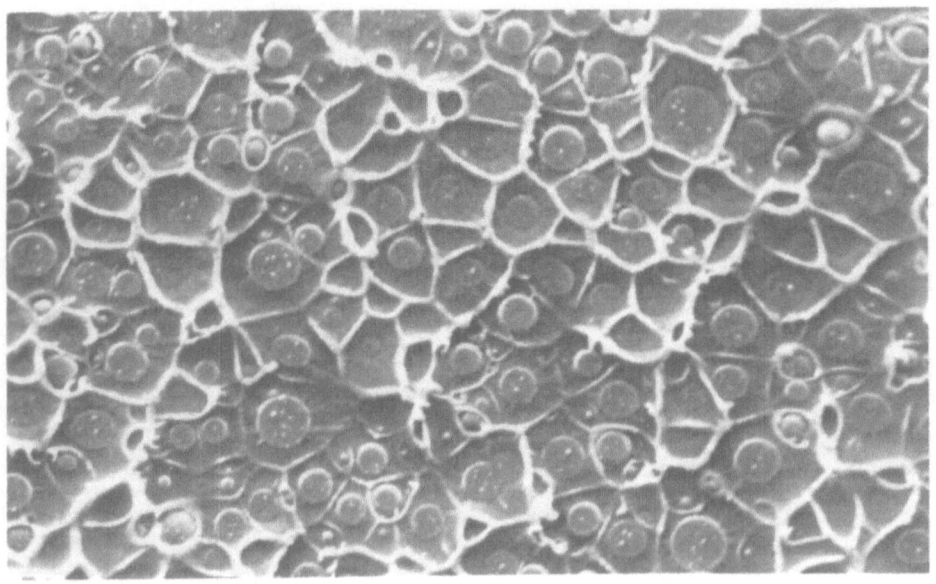

Fig. 9. PS/P(25MMA-75S) system with PS ($M_n$ = 15,100; W = 19.87 wt %) the dispersed phase and P(25MMA-75S) the continuous phase. 1980X.

Table I.   Summary of PS/PMMA Specimens (PS the Polymer Present in
the Initial PS/MMA Mixture) and Indication of Opacity[a]

| Wt % PS | Molecular weight PS | Opacity | Wt % PS | Molecular weight PS | Opacity |
|---|---|---|---|---|---|
| 4.91 | 2100 | cl. | 4.96 | 19,650 | trans. |
| 10.04 | 2100 | cl. | 10.10 | 19,650 | op. |
| 19.95 | 2100 | cl. | 12.72 | 19,650 | op. |
| 30.02 | 2100 | op. | 15.58 | 19,650 | op. |
| 35.30 | 2100 | op. | 17.65 | 19,650 | op. |
| 39.83 | 2100 | op. | 20.20 | 19,650 | op. |
|  |  |  | 22.33 | 19,650 | op. |
| 4.98 | 3100 | cl. | 25.45 | 19,650 | op. |
| 10.04 | 3100 | cl. | 27.45 | 19,650 | op. |
| 20.02 | 3100 | op. | 30.02 | 19,650 | op. |
| 30.27 | 3100 | op. |  |  |  |
| 32.87 | 3100 | op. | 5.03 | 49,000 | trans. |
| 35.04 | 3100 | op. | 7.58 | 49,000 | op. |
| 37.41 | 3100 | op. | 8.68 | 49,000 | op. |
| 40.17 | 3100 | op. | 10.07 | 49,000 | op. |
|  |  |  | 15.19 | 49,000 | op. |
| 5.04 | 9600 | trans. | 19.79 | 49,000 | op. |
| 10.11 | 9600 | trans. | 22.77 | 49,000 | op. |
| 12.14 | 9600 | op. | 25.33 | 49,000 | op. |
| 14.97 | 9600 | op. | 29.58 | 49,000 | op. |
| 17.77 | 9600 | op. |  |  |  |
| 20.31 | 9600 | op. |  |  |  |
| 22.55 | 9600 | op. |  |  |  |
| 25.31 | 9600 | op. |  |  |  |
| 27.46 | 9600 | op. |  |  |  |
| 30.08 | 9600 | op. |  |  |  |

[a] cl. = perfectly clear; trans. = translucent; op. = opaque.

nearly identical to Figures 1 through 5.

Type [2].  PS is the dispersed phase and PMMA or P(MMA-S) is the
continuous phase.  Typical of this type are the specimens shown in Figures 6, 7, and 9; as are also the
additional examples illustrated in Figures 10 through
13.  Generally the dispersed PS particles are more-or-
less spherical, although the particles tend toward
elipsoidal or more irregular shapes at higher values
of W and $M_n$ as seen in Figures 12 and 13.

Type [3].  PMMA or P(MMA-S) is the dispersed phase and PS is the
continuous phase.  Figures 14 and 15 are typical of
this type where the dispersed particles always tend

Table II. Summary of PS/P(MMA-S) Specimens (PS the Polymer Present in the Initial PS/MMA-S Mixture) and Indication of Opacity[a]

| Wt % PS[b] | Molecular weight PS | Copolymer Composition | | | |
|---|---|---|---|---|---|
| | | PMMA | P(75MMA-25S) | P(50MMA-50S) | P(25MMA-75S) |
| 5 | 2100 | cl. | cl. | cl. | cl. |
| 5 | 3100 | cl. | cl. | cl. | cl. |
| 5 | 9600 | (trans.)[c] | trans. | cl. | cl. |
| 5 | 12,000 | op. | trans. | cl. | cl. |
| 5 | 19,650 | (trans.) | trans. | trans. | cl. |
| 5 | 50,400 | (trans.) | trans. | trans. | trans. |
| 10 | 2100 | cl. | cl. | cl. | cl. |
| 10 | 3100 | trans.- | cl. | cl. | cl. |
| 10 | 9600 | (trans.) | op. | tra s. | cl. |
| 10 | 12,000 | op. | op. | trans. | trans.- |
| 10 | 19,650 | (op.) | op. | op. | trans.- |
| 10 | 50,400 | op. | op. | op. | trans. |
| 15 | 2100 | cl. | cl. | cl. | cl. |
| 15 | 3100 | trans. | trans. | cl. | cl. |
| 15 | 12,000 | op. | op. | op. | trans. |
| 15 | 18,900 | op. | op. | op. | op. |
| 15 | 50,400 | op. | op. | op. | op. |
| 20 | 2100 | trans.- | cl. | cl. | cl. |
| 20 | 3100 | op. | op. | trans.- | trans.- |
| 20 | 12,000 | op. | op. | op. | trans.- |
| 20 | 15,100 | op. | op. | op. | trans. |
| 20 | 18,900 | op. | op. | op. | trans. |
| 20 | 50,400 | op. | op. | op. | op. |
| 25 | 2100 | trans. | trans.- | cl. | cl. |
| 25 | 3100 | op. | op. | trans. | cl. |
| 25 | 12,000 | op. | op. | op. | trans. |
| 25 | 15,100 | op. | op. | op. | trans. |
| 25 | 18,900 | op. | op. | op. | trans. |
| 25 | 50,400 | op. | op. | op. | op. |
| 30 | 2100 | trans. | trans. | trans.- | cl. |
| 30 | 3100 | op. | op. | op. | trans.- |
| 30 | 12,000 | op. | op. | op. | trans. |
| 30 | 15,100 | op. | op. | op. | trans. |
| 30 | 50,400 | op. | op. | op. | op. |

[a] cl. = perfectly clear; trans.- = very slightly opalescent; trans. = translucent; op. = opaque.

[b] Values of weight percent PS are rounded to the nearest whole number; in no case does the actual value differ by more than ± 0.7%.

[c] Five values in parentheses taken from earlier work (see Table I).

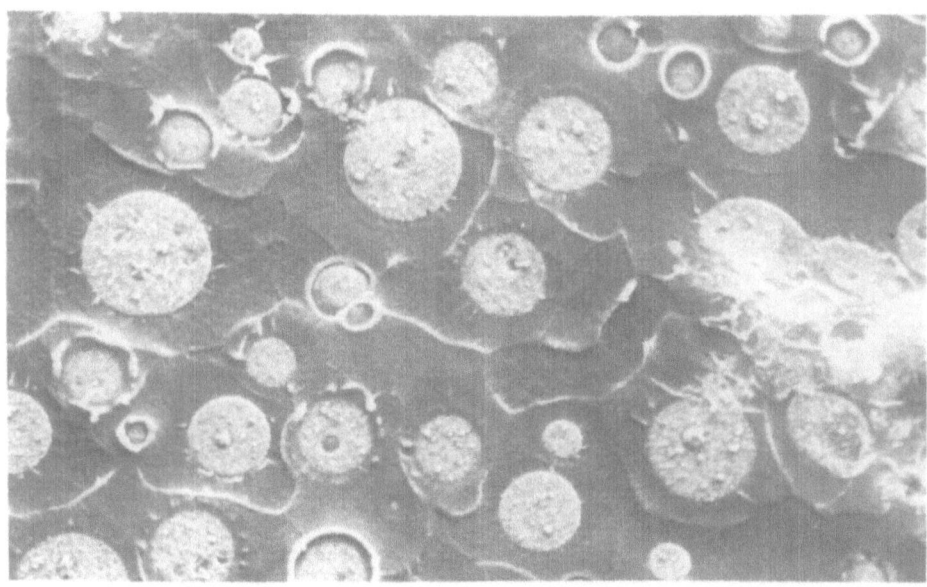

Fig. 10. PS/P(25MMA-75S) system with PS ($M_n$ = 50,400; W = 15.01 wt %)
the dispersed phase and P(25MMA-75S) the continuous phase.
1010X.

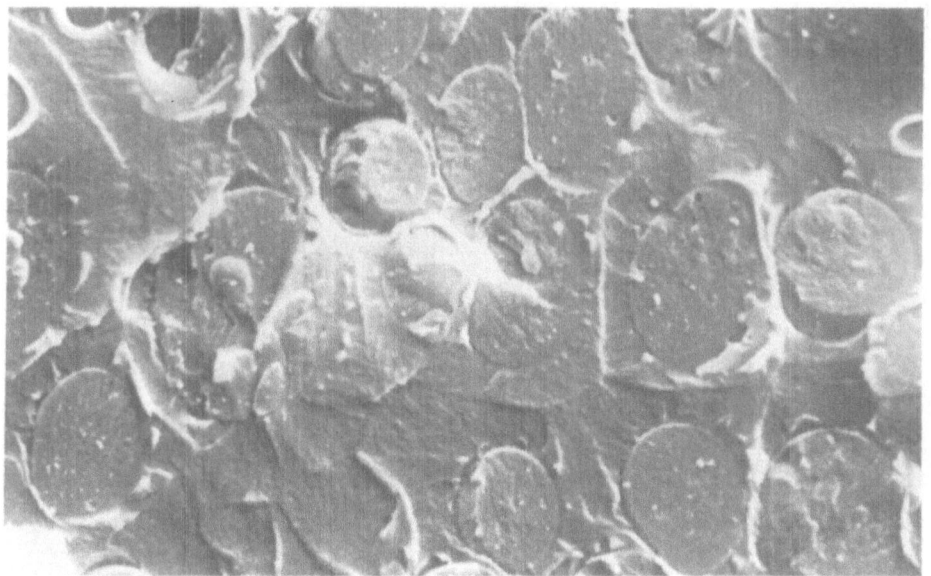

Fig. 11. PS/P(50MMA-50S) system with PS ($M_n$ = 12,000; W = 19.73 wt %)
the dispersed phase and P(50MMA-50S) the continuous phase
2125X.

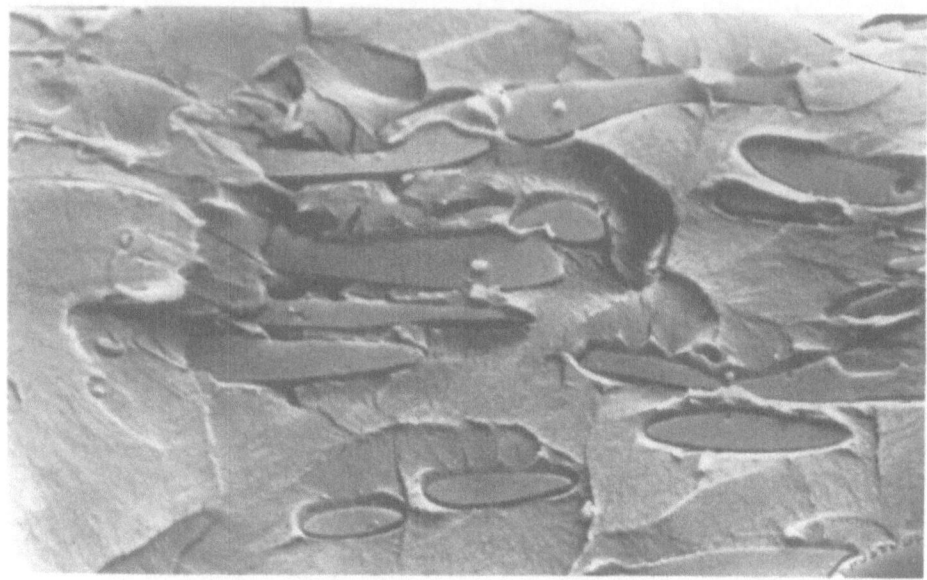

Fig. 12. PS/P(75MMA-25S) system with PS ($M_n$ = 12,000; W = 15.23 wt%)
the dispersed phase and P(75MMA-25S) the continuous phase.
2015X.

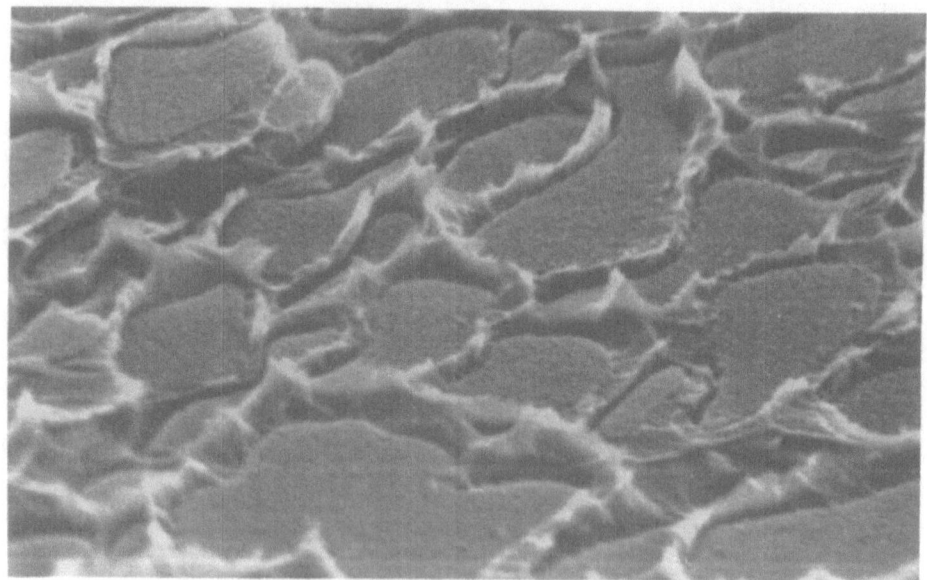

Fig. 13. PS/PMMA system with PS ($M_n$ = 3100; W = 30.27 wt %) the dispersed phase and PMMA the continuous phase.   4680X.

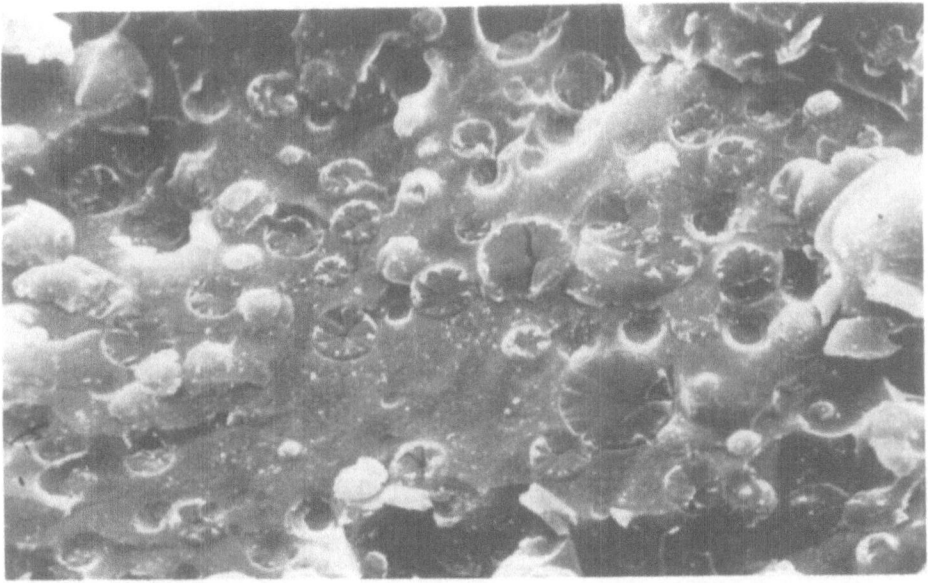

Fig. 14. PS/P(75MMA-25S) system with P(75MMA-25S) the dispersed
phase and PS ($M_n$ = 18,900; W = 25.17 wt %) the continuous
phase.   2160X.

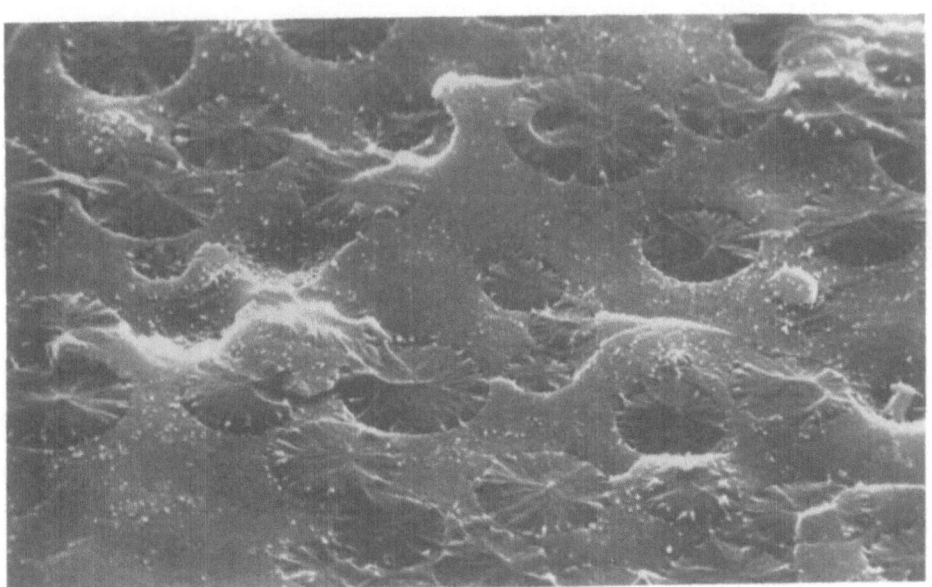

Fig. 15. PS/PMMA system with PMMA the dispersed phase and PS ($M_n$ =
19,650; W = 30.02 wt %) the continuous phase.   1600X.

to be spherical, or very nearly spherical, in form.
Type [4]. Specimens in which regions of PS dispersed in PMMA or
P(MMA-S) coexist with regions of PMMA or P(MMA-S) dispersed in PS. Very sharp boundaries separate the
regions which themselves have morphological characteristics in common with either type [2] or [3]. Figures
16, 17, and 18 represent typical examples of this
type.

In Tables III and IV are listed the values of the average size
of the dispersed particles ($\bar{S}$) for the PS/PMMA specimens from our
earlier work and the PS/P(MMA-S) specimens under consideration here,
respectively. The average size is defined, perhaps somewhat arbitrarily, as the average of the geometrical mean diameter of the
particles in a given representative area. Some care must be taken
to avoid obtaining underestimates of $\bar{S}$ (39); usually the standard
deviation of our values is within 5%, and never greater than 10%.
In Tables III and IV, unstarred values of $\bar{S}$ refer to type [2] where
PS forms the dispersed phase; starred values refer to type [3] where
PMMA or P(MMA-S) forms the dispersed phase; and pairs of values
refer to type [4] where both PS and PMMA or P(MMA-S) form the dispersed phases. Values of $\bar{S}$ = 0 refer to type [1] where no dispersed
particles are present; and values of $\bar{S}$ < 4 generally represent
specimens that are classified in Tables I and II as being trans-

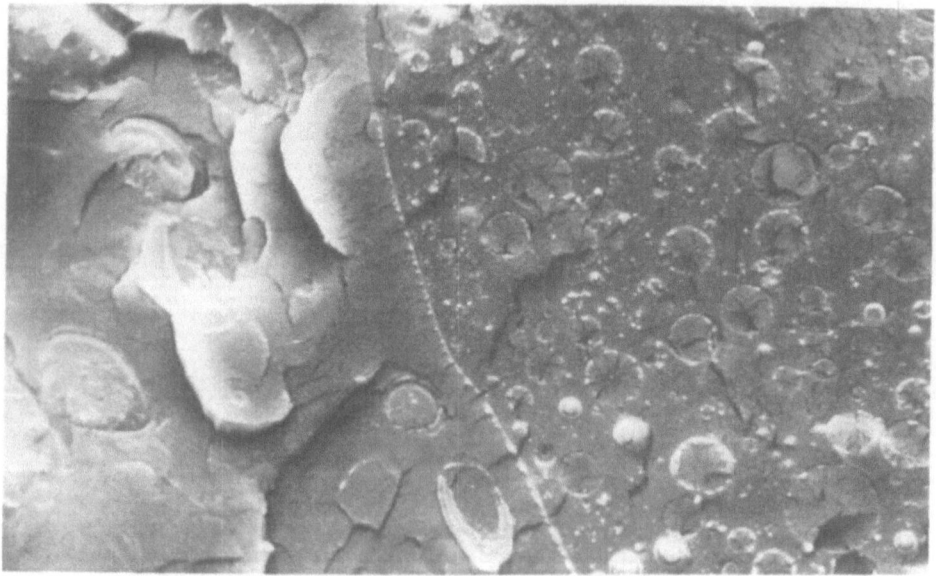

Fig. 16. PS/P(50MMA-50S) system with PS the dispersed phase in
P(50MMA-50S) (left side) and P(50MMA-50S) the dispersed
phase in PS (right side) ($M_n$ = 15,100; W = 24.79 wt %). 540X.

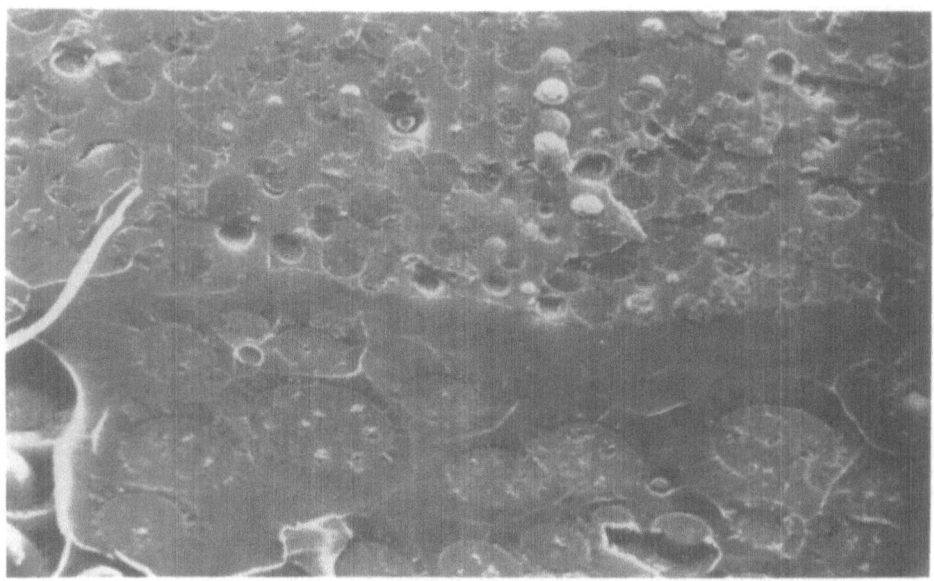

Fig. 17. PS/PMMA system with PS the dispersed phase in PMMA (lower part) and PMMA the dispersed phase in PS (upper part) ($M_n$ = 9600; W = 20.31 wt %).  425X.

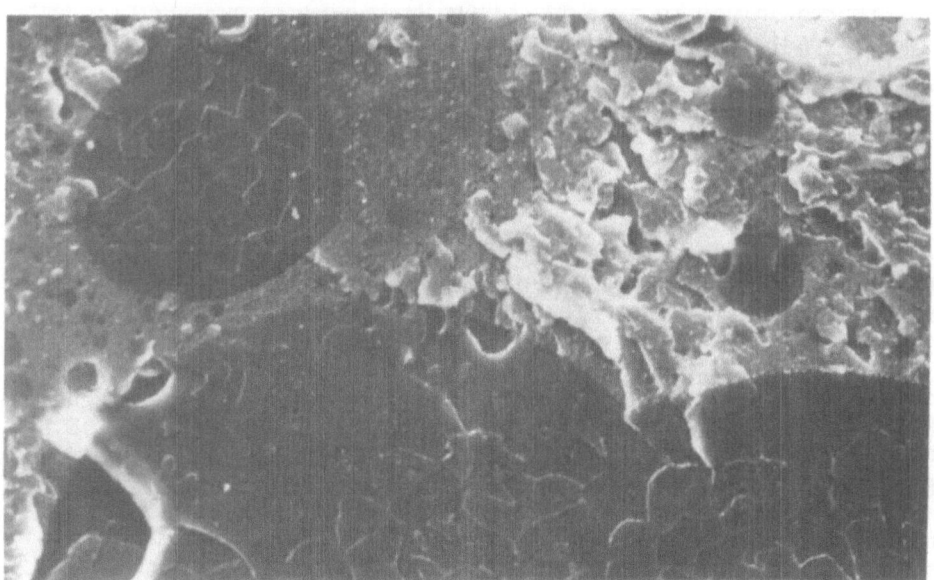

Fig. 18. PS/P(50MMA-50S) system with PS the dispersed phase in P(50MMA-50S) (upper part) and P(50MMA-50S) the dispersed phase in PS (lower part) ($M_n$ = 50,400; W = 30.00 wt %). 215X.

Table III. Dispersed Particle Size[a] ($\mu$m) as a Function of Weight
Percent PS and Number-Average Molecular Weight of PS
for the PS/PMMA System

| Wt % PS[b] | Number-average Molecular Weight PS | | | | |
|---|---|---|---|---|---|
|  | 2100 | 3100 | 9600 | 19,650 | 49,000 |
| 5.0 | 0 | 0 | 3.3 | 7.8 | 8.8 |
| 7.5 | --- | --- | --- | --- | 73.7 |
| 10.0 | 0 | 0 | 4.1 | 8.7 | 23.6/21.3* |
| 12.5 | --- | --- | 4.8 | 11.9 | --- |
| 15.0 | --- | --- | 5.5 | 17.6/15.2* | 9.6/20.0* |
| 17.5 | --- | --- | 9.6 | 18.3/14.0* | --- |
| 20.0 | 0 | 6.3 | 23.8/16.2* | 13.4/14.6* | 2.3/16.3* |
| 22.5 | --- | --- | 23.0/8.1* | 10.8/14.4* | 21.1* |
| 25.0 | --- | --- | 20.3/8.5* | 8.6* | 13.1* |
| 27.5 | --- | --- | 15.0* | 10.6* | --- |
| 30.0 | 1.0 | 6.6 | 10.0* | 2.2* | 6.8* |
| 32.5 | --- | 5.4/30* | --- | --- | --- |
| 35.0 | 1.9 | 12.9* | --- | --- | --- |
| 37.5 | --- | 9.7* | --- | --- | --- |
| 40.0 | 8.7 | 9.4* | --- | --- | --- |

[a] Unstarred values of $\bar{S}$ refer to type [2], PS dispersed in PMMA;
starred values to type [3], PMMA dispersed in PS; sets of values
to type [4], both PS dispersed in PMMA and PMMA dispersed in PS.
[b] Values of weight percent PS rounded off; see Table I for exact
values.

lucent. In Figures 19 through 22 we summarize the four types of
phase behavior discussed above, and indicate graphically their dependence on molecular weight and weight percent.

As we mentioned earlier, Massa has recently investigated the
question of compatibility of low-molecular-weight PS in PMMA and
P(MMA-S) (31). In terms of the present report, type [1] phase behavior represents a compatible polymer/polymer system, and there is
good qualitative agreement between the data reported here and that
obtained by massa. However, we do note that the limits of compatibility observed here for the PS/PMMA system (Figure 19) seem
somewhat less restrictive, while at the other extreme appear somewhat more restricted for the PS/P(25MMA-75S) system (Figure 22);
at present we have no systematic explanation to offer for these
rather minor differences.

Finally, we again draw attention to a comparison of Table III

Table IV. Dispersed Particle Size[a] ( μm) as a Function of Weight
Percent PS, Number-Average Molecular Weight PS, and Copolymer Composition for the PS/P(MMA-S) System

| Wt % PS[b] | Molecular Weight PS | Copolymer Composition | | | |
|---|---|---|---|---|---|
| | | PMMA | P(75MMA-25S) | P(50MMA-50S) | P(25MMA-75S) |
| 5 | 2100 | 0 | 0 | 0 | 0 |
| 5 | 3100 | 0 | 0 | 0 | 0 |
| 5 | 9600 | (3.3)[c] | 3.3 | 0 | 0 |
| 5 | 12,000 | 5.0 | 3.4 | 0 | 0 |
| 5 | 19,650 | (7.8) | 4.2 | 4.3 | 0 |
| 5 | 50,400 | (8.8) | 3.9 | 4.2 | 2.1 |
| 10 | 2100 | 0 | 0 | 0 | 0 |
| 10 | 3100 | 1.1 | 0 | 0 | 0 |
| 10 | 9600 | (4.1) | 4.5 | 1.6 | 0 |
| 10 | 12,000 | 5.0 | 5.0 | 2.7 | 0.5 |
| 10 | 19,650 | (8.7) | 7.0 | 4.8 | 0.7 |
| 10 | 50,400 | 20.0/15* | 19.9/3.9* | 7.0 | 4.2 |
| 15 | 2100 | 0 | 0 | 0 | 0 |
| 15 | 3100 | 2.0 | 2.8 | 0 | 0 |
| 15 | 12,000 | 10.2 | 7.2 | 6.8 | 0.9 |
| 15 | 18,900 | 16.4/17.0* | 7.1 | 7.7 | 5.1 |
| 15 | 50,400 | 9.1/14.6* | 9.4/3.2* | 15.0/6.3* | 31.5 |
| 20 | 2100 | 0.4 | 0 | 0 | 0 |
| 20 | 3100 | 5.1 | 4.4 | 0.5 | 0.5 |
| 20 | 12,000 | 9.9/7.3* | 12.5 | 7.5 | 0.5 |
| 20 | 15,100 | 16.0/15* | 11.5/10.7* | 12.4 | 2.1 |
| 20 | 18,900 | 13.0/17.6* | 9.4/38.0* | 10.7 | 2.3 |
| 20 | 50,400 | 1.5/16.2* | 4.0/10.4* | 1.8/21.5* | 39.2 |
| 25 | 2100 | 1.0 | 0.3 | 0 | 0 |
| 25 | 3100 | 5.9 | 3.0 | 2.1 | 0 |
| 25 | 12,000 | 15/8.0* | 0.3/19.1* | 7.4 | 1.0 |
| 25 | 15,100 | 15.1* | 6.5* | 9.1/6.3* | 3.0 |
| 25 | 18,900 | 8.6* | 6.1* | 2.5/5.8* | 2.1 |
| 25 | 50,400 | 11.3* | 8.3* | 0.5/13.7* | 42.5/5.0* |
| 30 | 2100 | 1.8 | 1.6 | 0.9 | 0 |
| 30 | 3100 | 12.6 | 4.1 | 4.2 | 0.2 |
| 30 | 12,000 | 9.0* | 7.9/2.6* | 7.4 | 1.1 |
| 30 | 15,100 | 6.9* | 8.1* | 9.8/5.0* | 2.0 |
| 30 | 50,400 | 6.8* | 9.8* | 1.8/5.3* | 7.9/4.9* |

[a] Unstarred values refer to type [2], PS dispersed in copolymer;
starred values to type [3], copolymer dispersed in PS; sets of
values to type [4], both PS dispersed in copolymer and copolymer
dispersed in PS.

[b] See footnote b to Table II.

[c] See footnote c to Table II.

and the PMMA column in Table IV, which represent many identical, or
very similar, specimens.  Inspection of these replica pairs indicates excellent agreement both with respect to the type of phase
behavior observed ([1], [2], [3], or [4]) and the quantitative
values of $\bar{S}$ obtained.  It was gratifying to us that acceptable reproducibility could be achieved given the number of variables involved in the polymerization of the samples.


DISCUSSION

    Four types of phase behavior characteristic of the PS/P(MMA-S)
system have been described and illustrated in some detail in the
previous section; and further, the average particle sizes have been
tabulated as a function of molecular weight and weight percent of
PS initially present in the PS/MMA-S mixture and of the composition
of the final P(MMA-S) copolymer resulting after polymerization.  In
the section we will discuss these results in terms of the ternary
polystyrene/poly(methyl methacrylate-styrene)/methyl methacrylate-
styrene phase diagram, dealing with (1) the four types of phase
relationships, (2) particle size, and lastly (3) multiple emulsions
or subinclusions within the dispersed phase.

    It should be noted that the following discussions involving
phase diagrams imply, strictly speaking, attainment of thermodynamic equilibrium during the polymerization process, that is,
all along the reaction path PS/MMA-S to PS/P(MMA-S).  This will
undoubtedly not be true for the polymerizations used to obtain our
test samples, especially after relatively high MMA-S to P(MMA-S)
conversions have been attained.  However, several interesting in-
sights manifest themselves as a result of analysis to be presented
in terms of phase diagrams and the properties of the polymerizing

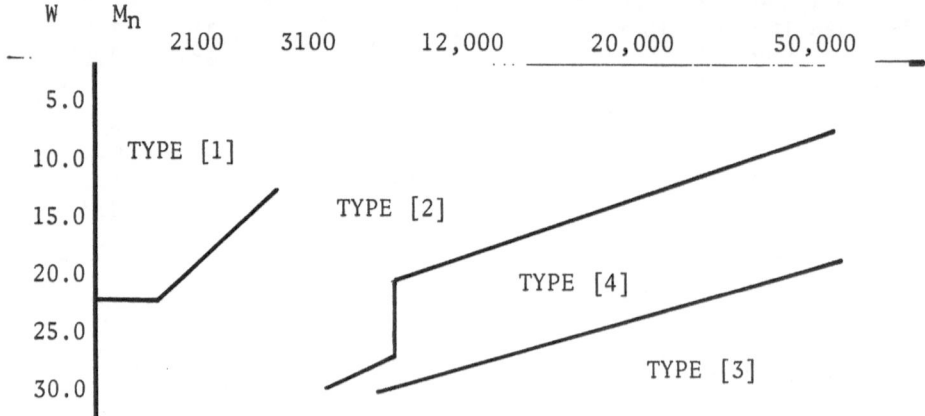

Fig. 19. Regions of four types of phase relationships observed
         in the PS/PMMA system.

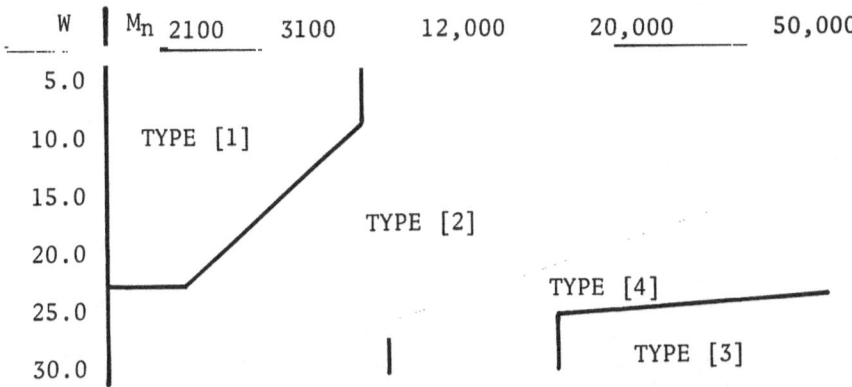

Fig. 20. Regions of four types of phase relationships observed in the PS/P(75MMA-25S) system.

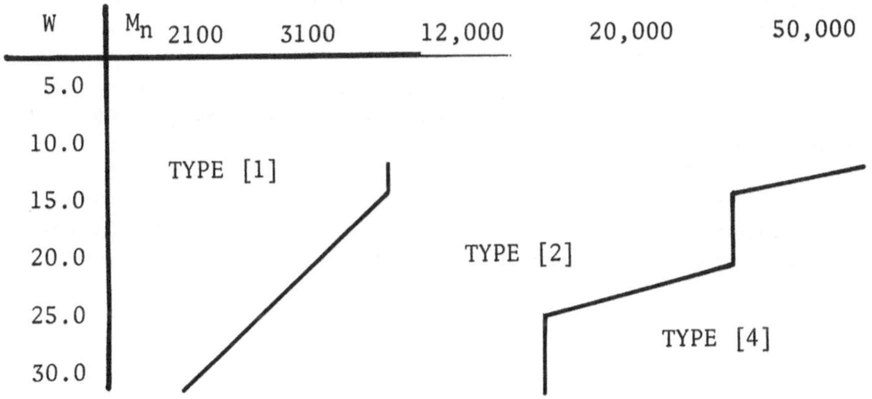

Fig. 21. Regions of three types of phase relationships observed in the PS/P(50MMA-50S) system.

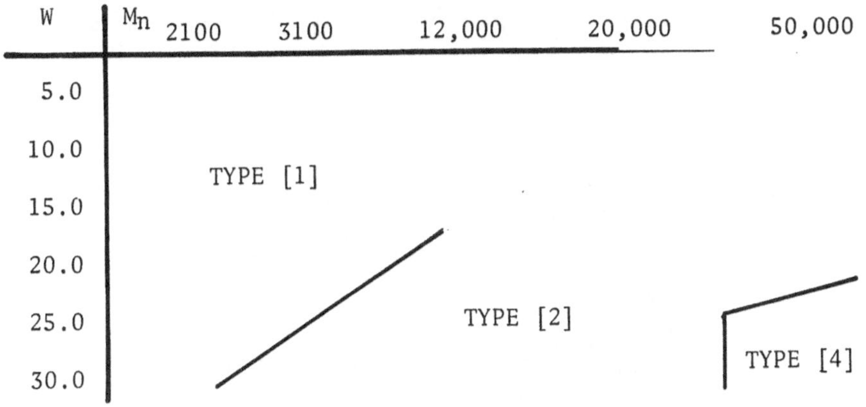

Fig. 22. Regions of three types of phase relationships observed in the PS/P(25MMA-75S) system.

system, and therefore seems to us to justify their use in the following discussions.

## Phase Relationships

There have been relatively few systematic investigations of the phase relationships of polymer/polymer/solvent systems, and of these we note in particular the studies of: Dobry and Boyer-Kawenoki (2) (polystyrene/polybutadiene/benzene), perhaps the first report to appear describing the phase diagram of a Polymer/polymer/solvent system; Kern and Slocombe (40) (a variety of vinyl polymers); and Paxton (41) (polystyrene/polybutadiene/toluene). Typically, the three-component phase behavior is as illustrated in Figure 23 for the polystyrene/polybutadiene/benzene system (2), where a homogeneous, one-phase solution of polystyrene and polybutadiene in benzene is separated by a phase boundary from a two-phase mixture consisting of a polystyrene-rich/polybutadiene phase and a polybutadiene-rich/polystyrene phase.  In common with similar three-component systems, there exists a critical point somewhere near the maximum of the phase boundary, and appropriate tie lines give the compositions and relative amounts of the two phases in the two-phase region.

The quantitative details of the low-molecular-weight polystyrene/poly(methyl methacrylate)/methyl methacrylate (hereafter denoted PS/PMMA/MMA) or the low-molecular-weight polystyrene/poly-

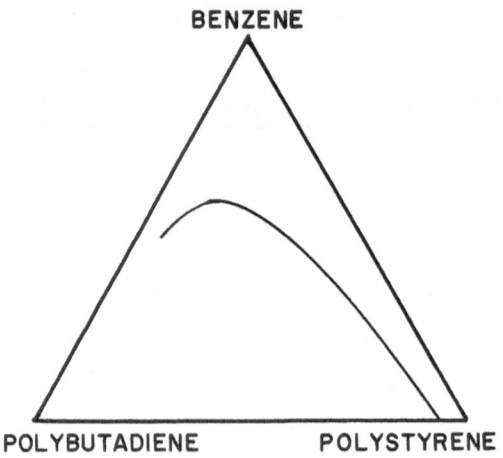

Fig. 23. Ternary phase diagram for the polystyrene/polybutadiene/benzene system (2).

(methyl methacrylate-styrene) copolymer/methyl methacrylate-styrene comonomer (hereafter denoted PS/P(MMA-S)/MMA-S) systems have not, to our knowledge, been worked out, particularly as a function of molecular weight of PS and P(MMA-S) composition. Nevertheless, based on the properties of similar systems, we assume phase diagrams and dependences on molecular weight of PS and P(MMA-S) composition as shown in Figures 24 and 25, respectively, where we note the following points:

1. As indicated in Figure 24, it is expected that the one-phase region is more restricted for higher-molecular-weight PS compared to lower-molecular-weight PS, as the mutual miscibility of two polymers in a common solvent is highly molecular weight dependent. Further, there probably is little dependence of the solubility of PMMA or P(MMA-S) on the molecular weight of PS in the vacinity of the PS apex since the molecular weight of PMMA or P(MMA-S) is assumed to be very high. On the other hand, in the vacinity of the PMMA or P(MMA-S) apex even minor changes in the molecular weight of PS will have a considerable influence on the mutual solubility since the molecular weight of PS can be as low as 2100.

2. As indicated in Figure 25, it is expected that the one-phase region is more restricted for higher MMA content compared to lower MMA content, as clearly in the limit of zero MMA content the system is miscible in all proportions. The behavior of the phase boundaries in the vacinity of the PS and PMMA or P(MMA-S) apexes is essentially the same as discussed above.

3. The critical point is not symmetrically located on the phase

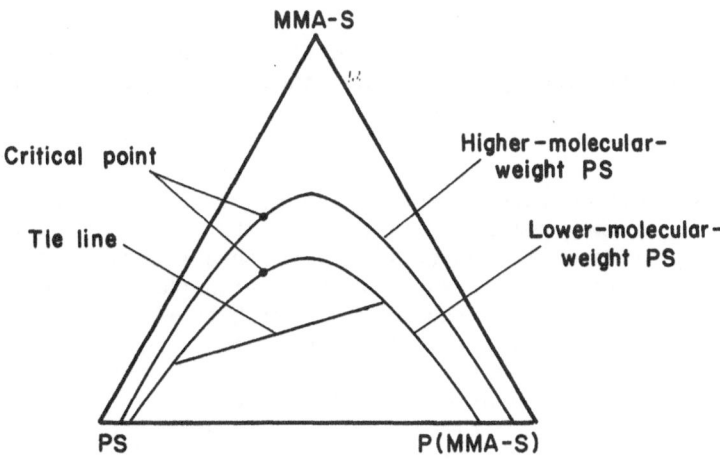

Fig. 24. Assumed ternary phase diagram for the PS/P(MMA-S)/MMA-S system for two molecular weights of PS.

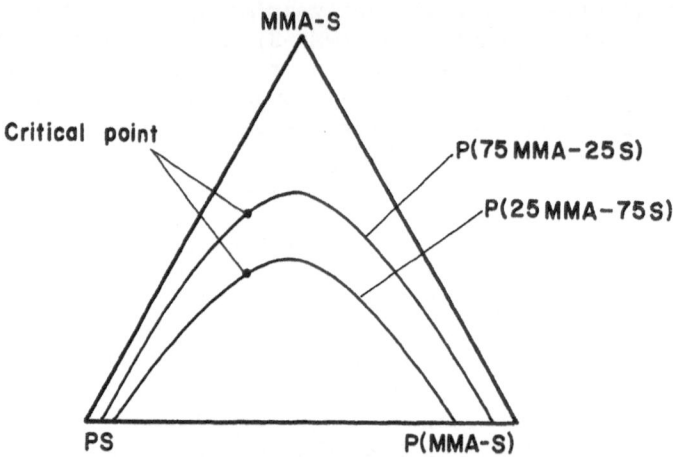

Pig. 25. Assumed ternary phase diagram for the PS/P(MMA-S)/MMA-S
system for two values of MMA content.

boundary but skewed towards the lower-molecular-weight polymer (42)
in this case towards the PS apex.

Turning now to the four types of phase behavior observed in
the PS/P(MMA-S) system (types [1], [2], [3], and [4] as defined in
the previous section), we refer to Figure 26 and summarize our
analysis in terms of the four reaction lines (hereafter denoted
RL) as follow:

Reaction line 1. The initial concentration of the PS/MMA-S
mixture is indicated by point a; and since RL1 never intersects
the phase boundary, a single phase solution of PS in P(MMA-S)
results. This corresponds to type [1] phase behavior.

Reaction line 2. The initial concentration is given by point b,
and phase separation takes place to the right of the critical point
at point c. Further polymerization along RL2 results in a two-
phase mixture of overall composition d; with a predominant P(MMA-S)-
rich phase of composition e and a lean PS-rich phase of composition
f, the relative amounts of the P(MMA-S)-rich to PS-rich phases
being df/de. When polymerization is complete, the final system
consists of a PS-rich phase dispersed in a  continuous, P(MMA-S)-
rich phase; and corresponds to type [2] behavior.

Reaction line 3. The initial concentration is given by point g,
and phase separation takes place to the left of the critical point
at point h. Further polymerization along RL3 results in a two-
phase mixture of overall composition k; with a prodominant PS-rich

phase of composition f and a lean P(MMA-S)-rich phase of composition
e, the relative amounts of the PS-rich to P(MMA-S)-rich phases
being ke/kf.  When polymerization is complete, the final system
consists of a P(MMA-S)-rich phase dispersed in a continuous, PS-rich
phase; and corresponds to type [3] behavior.

Reaction line 4.  The initial concentration is given by point m,
and phase separation takes place very near the critical point at
point n.  Further polymerization along RL4 results in a two-phase
mixture of overall composition p; with a P(MMA-S)-rich phase of
composition e and a PS-rich phase of composition f.  The ratio of
the tie lines, pf/pe, is close to unity and, hence, neither phase
tends to be predominant.  The final system consists of two more-or-
less continuous phases, one rich in P(MMA-S) with dispersed PS and
the other rich in PS with dispersed P(MMA-S); and corresponds to
type [4] phase behavior.

Examination of Tables III and IV indicates that higher molecular weight PS and/or higher weight percent PS and/or higher methyl
methacrylate content in the P(MMA-S) copolymer composition favors
the progression of phase behavior from type [1] to type [2] to
type [4] to type [3].  We will now examine these three variables
separately, discussing in order (1) weight percent PS; (2) molecular weight PS; and (3) MMA composition in P(MMA-S).

1. Weight percent PS.  Referring to Figure 26 again, we consider
four points of increasing weight percent PS (points a, b, m, and g)

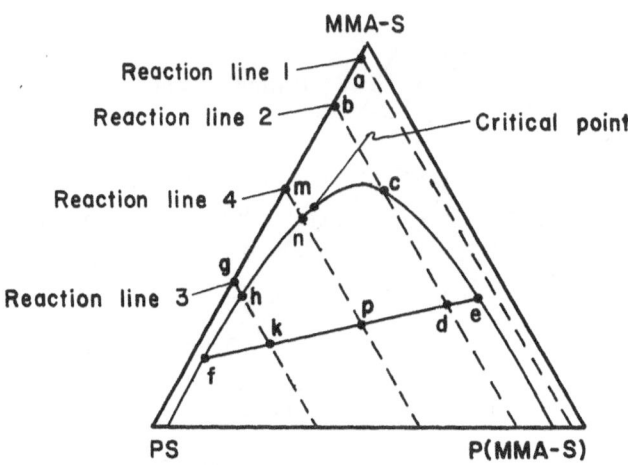

Fig. 26. PS/P(MMA-S)/MMA-S phase diagram with four reaction lines
        indicating various types of phase behavior.

for fixed values of molecular weight of PS and MMA content in
P(MMA-S). As we discussed above, points a, b, m, and g lead to
type [1], [2], [4], and [3] phase behavior, respectively. Examination of Tables III and IV shows numerous examples of trends of this
sort: for example, in Table III note the 3100 molecular weight
column; or in Table IV the $M_n$ = 50,400 and MMA = 75 wt % or the
$M_n$ = 12,000 and MMA = 50 wt % series.

2. Molecular weight PS. Referring to Figure 27 we consider two
different molecular weights of PS and fixed values of weight percent PS and MMA content in P(MMA-S), and make use of the analysis
advanced earlier in this subsection. For an initial PS/MMA-S composition given by point a, RL1 results in a type [2] system for the
higher-molecular-weight PS and a type [1] system for the lower-
molecular-weight PS. For an initial mixture given by point b, RL2
leads to a type [4] system for the higher-molecular-weight PS and
a type [2] system for the lower-molecular-weight PS. And finally,
for an initial mixture given by point c, RL3 results in a type [3]
system for the higher-molecular-weight PS and a type [4] system for
the lower-molecular-weight PS. These trends are in complete accord
with the data presented in Tables III and IV: for example, in
Table III note the 5 wt % row; and in Table IV the W = 10 wt % and
MMA = 75 wt % and the W = 25 wt % and MMA = 75 wt % series.

3. MMA content in copolymer. Referring to Figure 28 we consider
three different comonomer/copolymer compositions and fixed values of
weight percent and molecular weight PS, and again make use of the
analysis discussed above. For an initial PS/MMA-S composition given

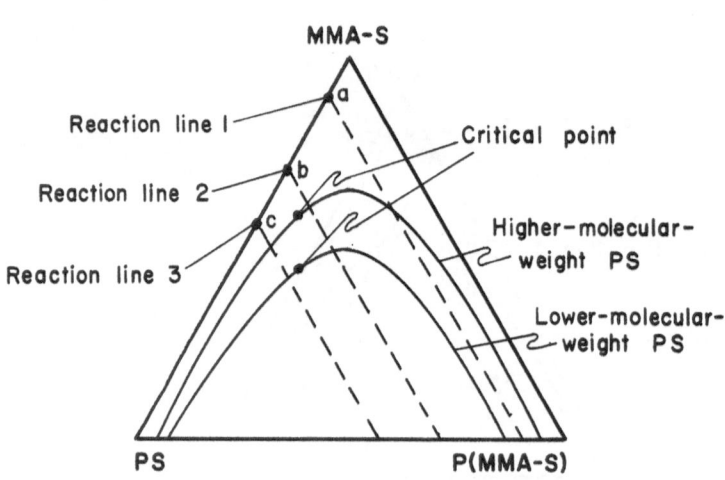

Fig. 27. PS/P(MMA-S)/MMA-S phase diagram with two reaction lines
         indicating dependence of two-phase behavior on molecular
         weight.

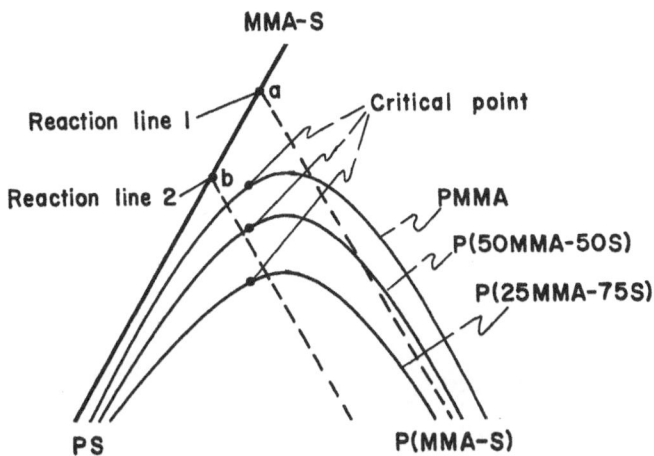

Fig. 28. PS/P(MMA-S)/MMA-S phase diagram with two reaction lines indicating dependence of two-phase behavior on MMA content.

by point a, RL1 results in a type [2] system for the PMMA and
P(50MMA-50S) cases and a type [1] system for the P(25MMA-75S) case.
Along RL2 for an initial mixture of composition b, we have the interesting situation where a type [3] system results for the PMMA
case, a type [4] system for the P(50MMA-50S) case, and a type [2]
system for the P(25MMA-75S) case.  Again these various trends are
in agreement with the data tabulated in Table IV.  In particular,
we note the series for W = 25 wt % and $M_n$ = 15,100 which displays
a type [3] to type [4] to type [2] progression with decreasing
MMA content; photomicrographs of this series are shown in Figures
29 through 31.

     To summarize, we note that the dependence of phase behavior,
as it manifests itself in the four types described and illustrated
in the preceeding section, on molecular weight and weight percent
PS and MMA content in the P(MMA-S) copolymer can be very satis-
factorily explained in terms of the ternary PS/P(MMA-S)/MMA-S phase
diagram.  It is apparent that a considerable degree of morphological control is possible through the variables being discussed
here; a striking example of this is seen in the sequence just
illustrated in Figures 29 through 31.

Dispersed Particle Size

     In the previous section we tabulated the values of average
particle size $\bar{S}$ obtained from our photomicrographs, and in this

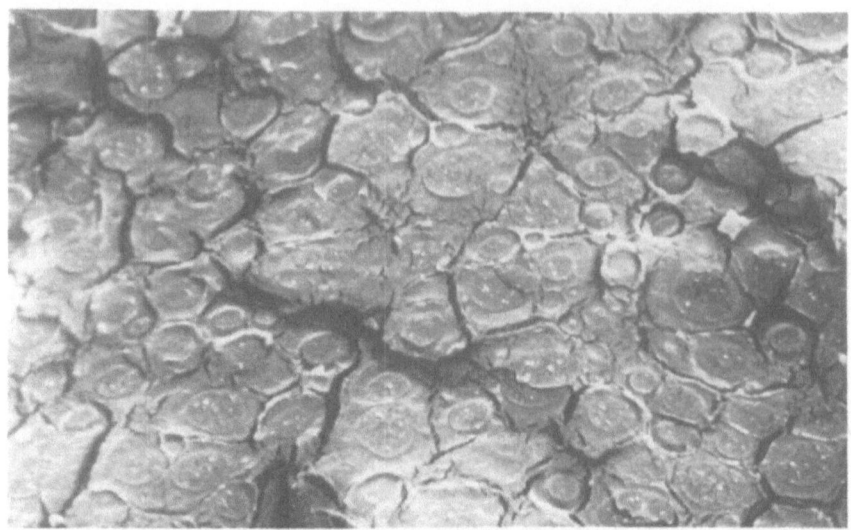

Fig. 29. PS/P(25MMA-75S) system with PS ($M_n$ = 15,100; W = 24.88 wt %)
the dispersed phase and P(25MMA-75S) the continuous phase.
2070X.

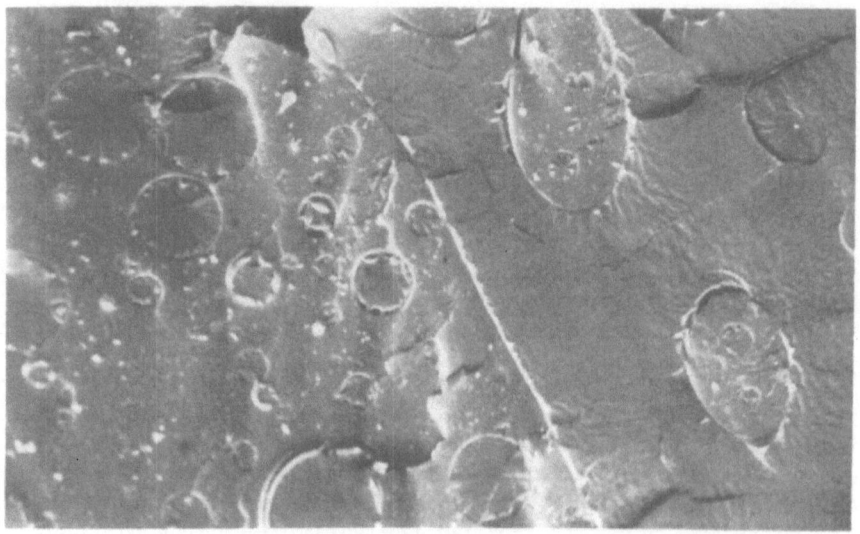

Fig. 30. PS/P(50MMA-50S) system with PS the dispersed phase in
P(50MMA-50S) (right side) and P(50MMA-50S) the dispersed
phase in PS (left Side) ($M_n$ = 15,100; W = 24.79 wt %).
1080X.

subsection we will proceed to discuss the trends in $\bar{S}$ as they relate
to the PS/P(MMA-S) phase diagram. From the point of view of the
quantitative relationship between $\bar{S}$ and the various types of phase
behavior (types [2], [3], and [4]), the PMMA and P(75MMA-25S) series
afford the most complete data, although analogous trends are seen
in the P(50MMA-50S) and P(25MMA-75S) series. The experimental
values of $\bar{S}$ reported in Table III are shown graphically in Figures
32 and 33; where PS and PMMA are the dispersed phases, respectively,
and the regions to the right of the single bar in Figure 32 and the
left of the double bar in Figure 33 are those where PS and PMMA
both coexist as the continuous phases. In Figure 34 is shown similar data for the P(75MMA-25S) system where PS is the dispersed
phase, and the region to the right of the single bar is that in which
PS and P(75MMA-25S) coexist as the continuous phases. Finally, to
illustrate the dependence of $\bar{S}$ on W and $M_n$ of PS, in Figure 35 is
shown data for $M_n$ = 50,400. It should be noted that in Figures
34 and 35, the data are insufficient to define the maximum values
of $\bar{S}$, and the illustrated peaks reflect this.

Examination of Tables III and IV and Figures 32 through 35
leads to the following conclusions concerning the dependence of
$\bar{S}$ on $M_n$ and W of PS and MMA content in P(MMA-S), namely:

1. For $M_n$ and MMA content constant, when PS is the dispersed
phase $\bar{S}$ increases with W in the type [2] region and then decreases

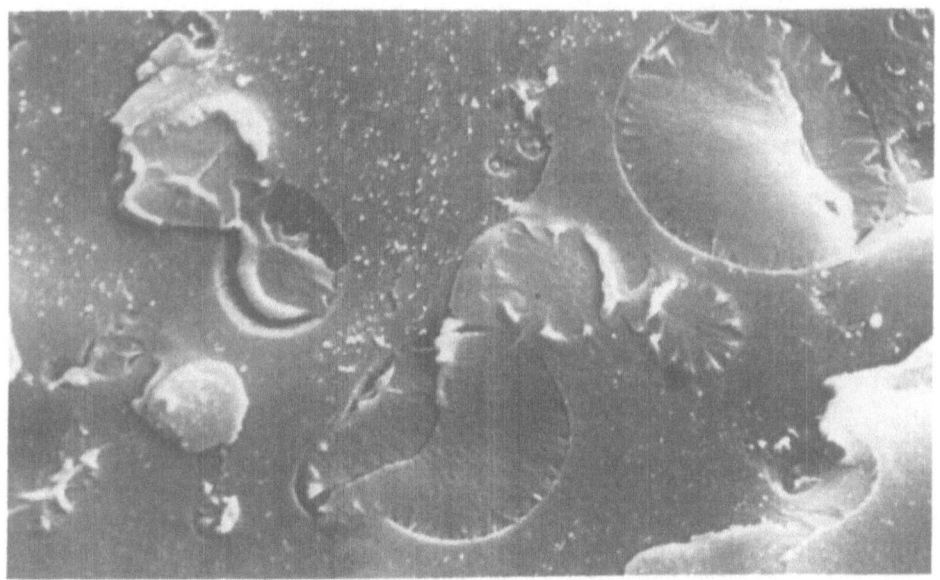

Fig. 31. PS/P(75MMA-25S) system with P(75MMA-25S) the dispersed
        phase and PS ($M_n$ = 15,100; W = 24.99 wt %) the continuous
        phase. 2125X.

rapidly once the type [4] region is entered.

2.  For W and MMA content constant, when PS is the dispersed phase $\bar{S}$ increases with $M_n$ in the type [2] region of phase behavior.

3.  For $M_n$ and W constant, when PS is the dispersed phase $\bar{S}$ decreases with decreasing MMA content in P(MMA-S) in the type [2] region.

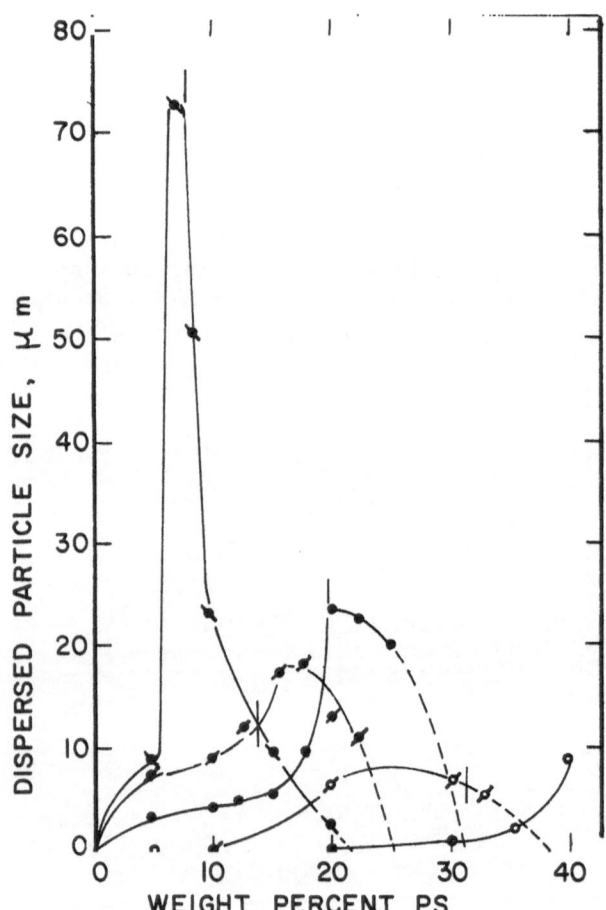

Fig. 32. Dispersed particle size versus weight percent PS for various values of number-average molecular weight PS for the
PS/PMMA system.  PS is the dispersed phase; and the region
in which both PS and PMMA coexist as the continuous phases
is to the right of the single bar.  o: $M_n$ = 2100; $\varnothing$: $M_n$ =
3100;  •: $M_n$ = 9600;  ♪: $M_n$ = 19,650;  ➤: $M_n$ = 49,000.

4.  Although sufficient data is available only for the PMMA case
for $M_n$ = 19,650 and 49,000, when PMMA is the dispersed phase $\bar{S}$ is
more-or-less constant in the type [4] region and then decreases
when the type [3] region is entered.

We will now proceed to examine these four trends in terms of the
PS/P(MMA-S)/MMA-S phase diagram, and certain assumptions concerning
the properties of the polymerizing system at the point of phase separation.  In particular, we assume that at the point of phase
separation and beyond, the dispersed particles in a less viscous
medium have a higher coalescence rate and therefore ultimately
form larger particles.

Turning first to the case where PS forms the dispersed phase
(type [2] behavior), we refer to Figure 36.  Comparing two values
of W for constant values of $M_n$ and MMA content, we note that along
RL1 (the lower value of W) phase separation occurs at point a
whereas along RL2 (the higher value of W) phase separation occurs

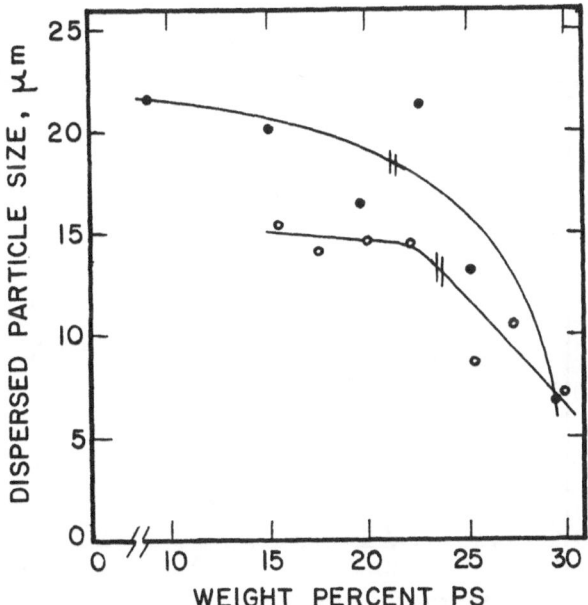

Fig. 33. Dispersed particle size versus weight percent PS for two
         values of number-average molecular weight of PS for the PS/
         PMMA system.  PMMA is the dispersed phase; and the region
         in which both PS and PMMA coexist as the continuous phases
         is to the left  of the double bar.  o: $M_n$ = 19,650;
         •: $M_n$ = 49,000.

at point b which represents a much lower degree of conversion of
monomer to polymer. Therefore, along RL2 phase separation occurs
in a less viscous medium than along RL1 (assuming that the increased initial viscosity due to somewhat higher weight percent
PS does not offset this effect due to decreased monomer conversion),
and results in larger particle size for the higher value of W.

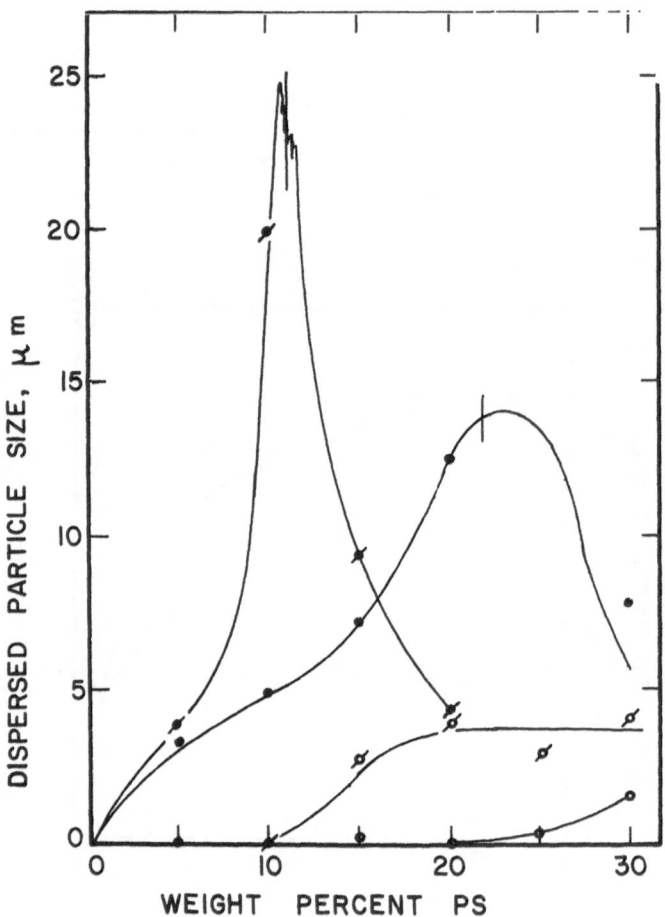

Fig. 34. Dispersed particle size versus weight percent PS for various values of number-average molecular weight PS for the
P(75MMA-25S) system. PS is the dispersed phase; and the
region in which both PS and P(75MMA-25S) coexist as the
continuous phases is to the right of the single bar.
o: $M_n$ = 2100; ∅: $M_n$ = 3100; •: $M_n$ = 12,000; ∅: $M_n$ = 50,400.

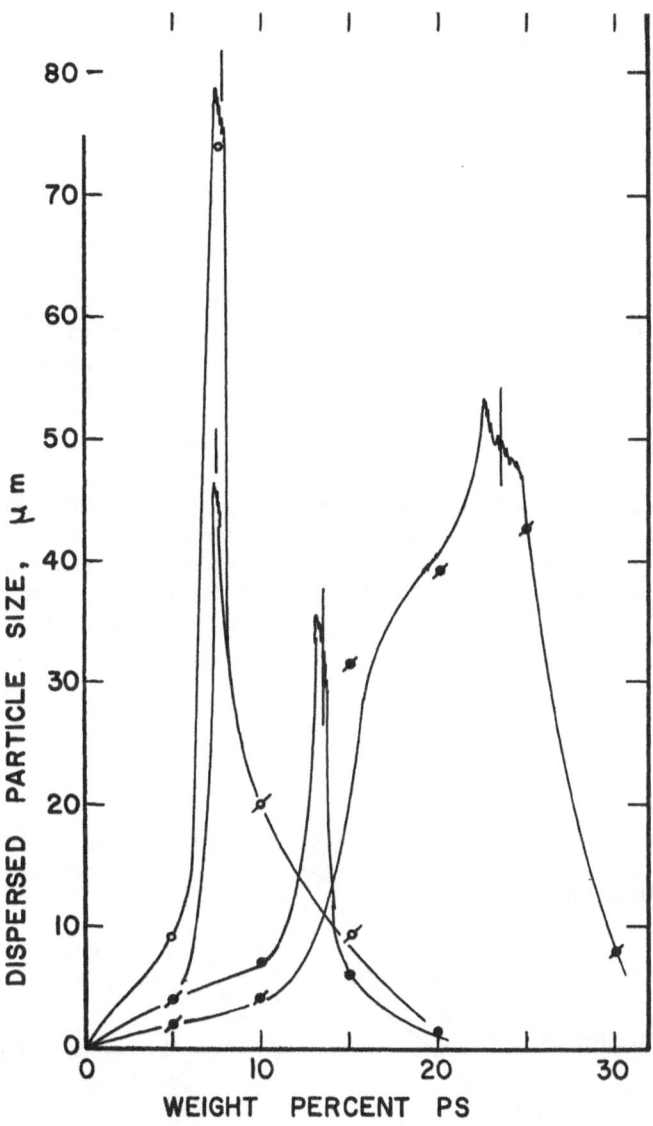

Fig. 35. Dispersed particle size versus weight percent PS for
$M_n$ = 50,400 and several P(MMA-S) copolymer compositions.
PS is the dispersed phase ; and the region in which both
PS and P(MMA-S) coexist as the continuous phases is to
the right of the single line.  o: PMMA; ◐: P(75MMA-25S);
●: P(50MMA-50S); ◗: P(25MMA-75S).

Similarly, considering the effect of two values of $M_n$ for constant
W and MMA content, we note that along RL2  phase separation occurs
at point b for higher $M_n$ and at point c for lower $M_n$.  Because of
decreased monomer-to-polymer conversion at point b compared to
point c, phase separation occurs in a less viscous medium for the
higher value of $M_n$ than the lower value of $M_n$ (assuming that the
increased initial viscosity due to higher $M_n$ does not offset this
effect due to decreased monomer conversion), and results in larger
particle size for the higher value of $M_n$.  Finally, comparing two
values of MMA content in P(MMA-S) for constant values of $M_n$ and W,
we refer to Figure 37.  Along RL1 phase separation for the PMMA case
occurs at point a and the P(50MMA-50S) case at point b.  Because of
the decreased monomer-to-polymer conversion at point a compared to
point b, phase separation occurs in a less viscous medium for the
PMMA case compared to the P(50MMA-50S) case (assuming that differences in the initial viscosity due to differences in PMMA and
P(50MMA-50S) do not offset  this effect due to decreased monomer
conversion), and result in larger particle size with increased MMA
content in P(MMA-S).  These three  conclusions are in accord with
the experimental observations seen in Figures 32, 34, and 35.

Considering now the situation when P(MMA-S) forms the dispersed
phase (type [3] behavior) we have sufficient data, as shown in
Figure 33, to analysis only the PS/PMMA case.  We refer again to
Figure 36, first comparing two values of $M_n$ for a given value of W.
We note that along RL3 phase separation occurs at point d for the

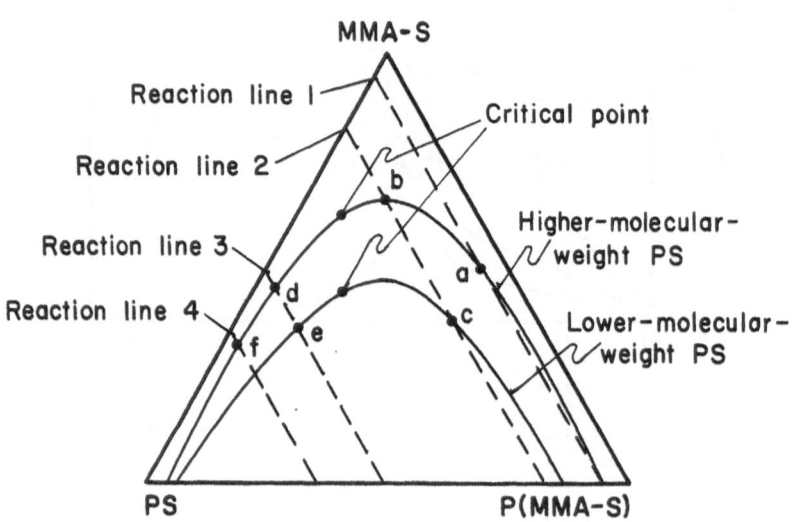

Fig. 36. PS/P(MMA-S)/MMA-S phase diagram indicating relationships
         between molecular weight and weight percent PS and particle
         size.

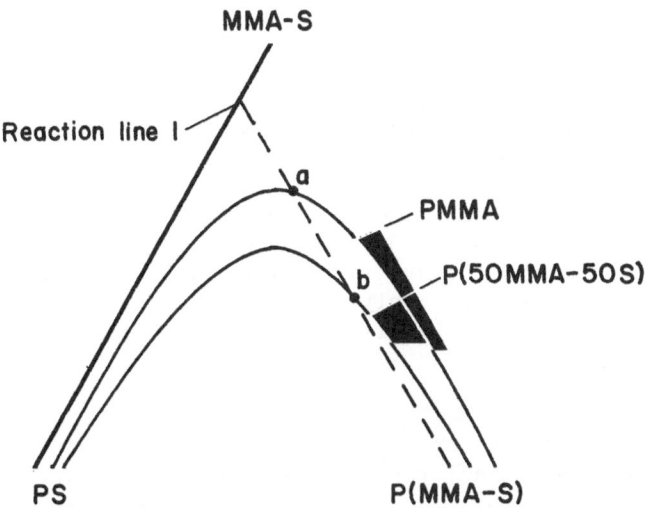

Fig. 37. PS/P(MMA-S)/MMA-S phase diagram indicating relationships
between weight percent PS and MMA content in copolymer,
and particle size.


higher value of $M_n$ and point e for the lower value of Mn.  Since
point d represents a less viscous medium compared to point e, higher
values of $M_n$ lead to larger particle sizes, in exact analogy to the
type [2] behavior described above when PS forms the dispersed phase.
However, comparing now two values of W for a constant value of $M_n$,
we find a distinct difference between type [2] and type [3] behavior.
For this latter case phase separation occurs at points d and f along
RL3 and 4, respectively; and since the phase boundary is roughly
parallel to the PS-MMA edge of the triangular diagram, the degree
of conversion represented by points d and f will be nearly equal.
Therefore, we expect, because of its higher initial concentration
of PS, that the system at point f will be somewhat more viscous
than that at point d; and that in the type [3] region higher values
of W for a given value of $M_n$ lead to decreased particle size.  The
two conclusions  discussed above are in agreement with the trends
shown in Figure 33.

     Finally, considering the region of type [4] behavior, we find
a somewhat more complex dependence of $\bar{S}$ on W, $M_n$, and, to the
extent that the data allow analysis, MMA content.  Two trends seem
to emerge  from the data, namely: (1) when PS is the dispersed
phase there is a rapid decrease in $\bar{S}$ with increasing W, and (2)
when PMMA is the dispersed phase there is relatively little dependence of $\bar{S}$ with increasing W and a small increase of $\bar{S}$ with increasing $M_n$.  Considering point (1) we refer to Figure 38 and note
that as W increases from RL1 to RL2, the relative amount of the

P(MMA-S)-rich phase decreases, that is, the relative amount of
db/dc (RL2) is less than da/dc (RL1).  Thus, the effective amount
of PS present in the P(MMA-S)-rich phase decreases due to its com-
version to the PS-rich continuous phase.  Thus, fewer dispersed
particles of PS form, lowering the likelihood of coalescence
occuring.  As far as point (2) is concerned, we can only note that
the data suggest a trade-off with viscosity, that is, the increase in initial viscosity due to increasing W is balanced by the
decrease in viscosity resulting from phase separation taking
place at a lower overall conversion.  Obviously, the discussion
advanced above can be taken only as a partial explanation of the
observed results, as the overall phenomena are complicated.

Multiple Emulsions

     Multiple emulsions, or subinclusions, were observed in many of
the PS/P(MMA-S) specimens; for example, they are clearly visible in
Figures 6, 8, 9, 10, 16, 17, and 30, and a spectacular example is
illustrated in Figure 39.  This phenomenon can  be understood in
terms of the PS/P(MMA-S) phase diagram and the viscosity of the
polymerizing system, and we conclude our report with this discussion.

     Referring to Figure 40, consider a system with an initial
PS/MMA-S composition given by point a and polymerizing along RL1.
Phase separation takes place at point b and at some later time the
system reaches an overall composition represented by point c, with

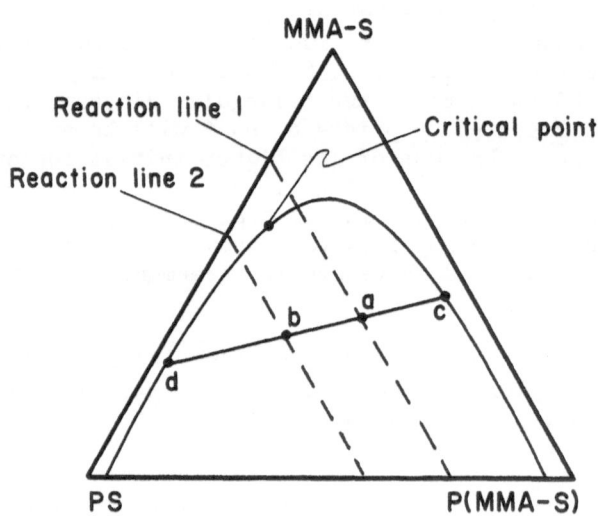

Fig. 38. PS/P(MMA-S)/MMA-S phase diagram indicating type [4]
         behavior.

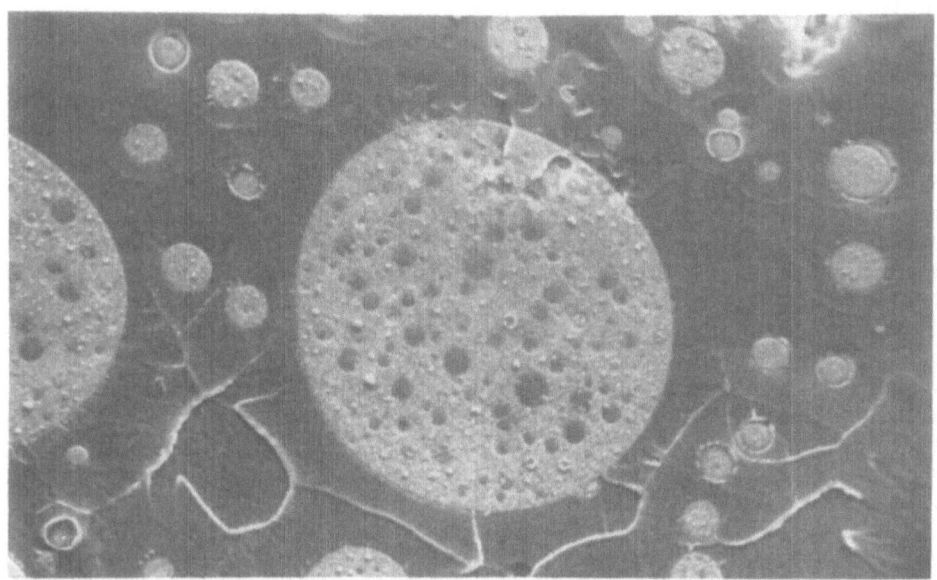

Fig. 39. PS/P(25MMA-75S) system with PS ($M_n$ = 50,400; W = 15.31 wt %) the dispersed phase with multiple emulsions and P(25MMA-75S) the continuous phase.    510X.

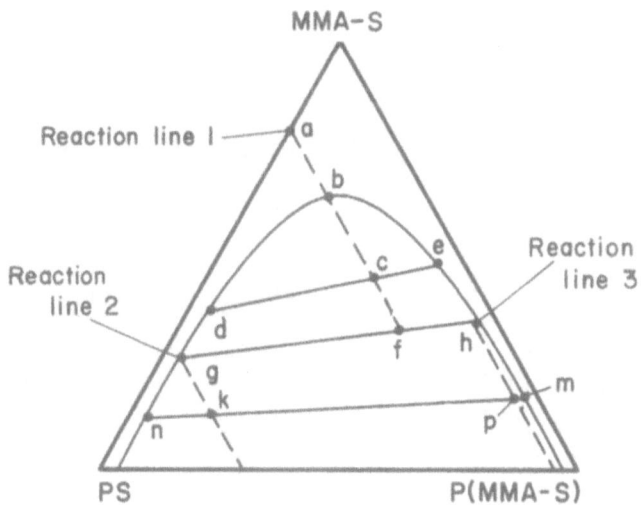

Fig. 40. PS/P(MMA-S)/MMA-S phase diagram indicating reaction lines leading to multiple emulsions.

a PS-rich phase of composition d and a P(MMA-S)-rich phase of composition e in the realtive amounts ce/cd. As polymerization continues, the viscosity of the system increases and at some point it seems reasonable to suppose that the viscosity becomes so high that the two phases become essentially isolated from each other. If this point corresponds to an overall composition given by point f, then the system consists of two isolated phases, a PS-rich phase of composition g and a P(MMA-S)-rich phase of composition h, in relative amounts fh/fg. Further polymerization of the PS-rich phase will now proceed along RL2 and results, at point k for example, in a phase rich in P(MMA-S) of composition m (the multiple emulsions) dispersed in the PS-rich phase of composition n. Similarly, further polymerization of the original continuous phase will continue along RL3, and result, at point p for example, in a PS-rich phase dispersed in a continuous P(MMA-S)-rich phase. Multiple emulsions within the continuous phase occur rather infrequently, certainly as compared to the occurrance of multiple emulsions within dispersed phases; however, we have observed several examples and these will be discussed in a future report.

REFERENCES

1. See, for example, M. W. Beijerinck, Kolloid Z. Z. Polym. 7:16 (1910).
2. See, for example, A. Dobry and F. Boyer·Kawenoki, J. Polym. Sci. 2:90 (1947).
3. N. A. J. Platzer, ed., 1971, "Multicomponent Polymer Systems" (Adv. Chem. Ser. 99), American Chemical Society, Washington, D.C.
4. L. H. Sperling, ed., 1974, "Recent Advances In Polymer Blends, Grafts, and Blocks," Plenum, New York.
5. N. A. J. Platzer, ed., 1975, "Copolymers, Polyblends, and Composites" (Adv. Chem. Ser. 142), American Chemical Society, Washington, D.C.
6. J. A. Manson and L. H. Sperling, 1976, "Polymer Blends and Composites," Plenum, New York.
7. D. R. Paul and S. Newman, eds., 1978, "Polymer Blends," Vols. I and II, Academic Press, New York.
8. S. L. Cooper and G. M. Estes, eds., 1979, "Multiphase Polymers" (Adv. Chem. Ser. 176), Americal Chemical Society, Washington, D. C.
9. G. C. Claver, Jr. and E. H. Merz, Off. Dig. Fed. Paint Varn. Prod. Clubs, 28:858 (1956).
10. P. A. Traylor, Anal. Chem., 33:1629 (1961).
11. J. Mann, R. J. Bird, and G. Rooney, Macromol. Chem., 90:207 (1966).
12. H. Keskkula and P. A. Traylor, J. Appl. Polym. Sci., 11:2361 (1967).
13. K. Kato, Jpn. Plastics, 2(2):6 (1968).
14. M. Matsuo, Jpn. Plastics, 2(3):6 (1968).

15. R. N. Haward and I. Brough, Polymer, 10:724 (1969).
16. H. Keskkula, Appl. Polym. Symp., 15:51 (1970).
17. R. J. Williams and R. W. A. Hudson, Polymer, 8:643 (1967).
18. C. B. Bucknall and R. R. Smith, Polymer, 6:437 (1965).
19. M. Matsuo, Polym. Eng. Sci., 9:206 (1969).
20. M. Baer, J. Appl. Polym. Sci., 16:1109 (1972).
21. J. D. Moore, Polymer, 12:478 (1971).
22. G. E. Molau, J. Polym. Sci., Part A, 3:1267 (1965).
23. Ibid., 4235.
24. G. E. Molau, J. Polym. Sci., Part B, 3:1007 (1965).
25. G. E. Molau and H. Keskkula, J. Polym. Sci., Part A-1, 4:1595 (1966).
26. G. E. Molau, W. M. Wittbrodt, and V. E. Meyer, J. Polym. Sci., 13:2735 (1969).
27. G. E. Molau, Kolloid Z. Z. Polym., 238:493 (1970).
28. B. W. Bender, J. Appl. Polym. Sci., 9:2887 (1965).
29. R. L. Kruse, in Ref. 5 above, pp. 141-147.
30. For a recent review, see D. R. Paul and J. W. Barlow, in Ref. 8 above, pp. 315-335.
31. D. J. Massa, in Ref. 8 above, pp. 433-442.
32. R. R. Parent and E. V. Thompson, Polym. Prepr., 18(2):507 (1977).
33. R. R. Parent and E. V. Thompson, J. Polym. Sci., Polymer Physics Edition, 16:1829 (1978).
34. R. R. Parent and E. V, Thompson, Polym. Prepr., 19(1):180 (1978).
35. R. R. Parent and E. V. Thompson, in Ref. 8 above, pp. 381-411.
36. E. V. Thompson, Org. Coat. Plast. Chem., 40:751 (1979).
37. See, for example, (a) J. P. Berry, J. Polym. Sci., Part C, 3:91 (1963); (b) R. N. Haward and J. Mann, Proc. R. Soc. London Ser. A, 282:120 (1964); (c) M. J. Doyle, A. Maranci, E. Orowan, F. R. S. Stork and S. T. Stork, ibid., 329:137 (1972); and (d) P. Beahan, M. Bevis, and D. Hull, Polymer, 14:96 (1973).
38. See, for example, (a) A. Chapiro, 1962, "Radiation Chemistry of Polymeric Systems," Interscience, New York, pp. 509-512; and (b) E. V. Thompson, J. Polym. Sci., Part B, 3:675 (1965).
39. For a discussion of this point, see (a) Ref. 21 and (b) D. M. Schwartz, J. Microsc., 96(1):25 (1972).
40. R. J. Kern and R. J. Slocombe, J. Polym. Sci., 15:183 (1955).
41. T. R. Paxton, J. Appl. Polym. Sci., 7:1499 (1963).
42. R. L. Scott, J. Chem. Phys., 17:279 (1949).

PREPARATION OF HIGHLY BRANCHED GRAFT COPOLYMERS BY

CHAIN TRANSFER REACTION

Shigeo Nakamura, Hideo Kasatani, and Kei Matsuzaki

Department of Industrial Chemistry, Faculty of Engineering, University of Tokyo, Bunkyo-ku, Tokyo, Japan

INTRODUCTION

Aromatic nitro compounds act as inhibitors or retarders in the radical polymerization of vinyl monomers. By the reaction of the aromatic nitro compounds with growing polymer radicals, the formation of addition products has been established between growing polymer radicals and the oxygen atoms of the nitro groups (1). Accordingly, a vinyl monomer is polymerized with a radical initiator in the presence of a polymer with pendant aromatic nitro groups, then a graft copolymer is obtained through the chain transfer reaction of growing polymer radicals to the aromatic nitro groups.

As trunk polymers, we used several polymers having pendant $p$-nitrophenyl groups such as poly(vinyl $p$-nitrobenzoate), poly($p$-nitrophenyl acrylate) and ethylene-vinyl $p$-nitrobenzoate copolymer, and carried out graft copolymerization of styrene onto these trunk polymers.

The reason why we used these polymers as trunk polymers is that the grafted side chains are easily isolated from the trunk polymer by hydrolysis of ester bonds, and characterization of grafted branches is possible.

EXPERIMENTAL

Poly(vinyl $p$-nitrobenzoate) was prepared by the esterification of poly(vinyl alcohol) with $p$-nitrobenzoyl chloride (2). Poly($p$-nitrophenyl acrylate) was obtained by bulk-polymerization of $p$-nitrophenyl acrylate (3). Ethylene-vinyl $p$-nitrobenzoate copolymer

containing 28.6 mole% of vinyl $p$-nitrobenzoate was prepared by
the esterification of ethylene-vinyl alcohol copolymer.

    Graft copolymerization was carried out at 60°C in evacuated
sealed ampoules using styrene as a monomer, azobisisobutyronitrile
(AIBN) as an initiator and dimethylformamide(DMF) as a solvent.
After a predetermined period of reaction, the contents were poured
into a mixture of methanol and water, and the precipitate was
filtered and dried.  Nongrafted polystyrene was removed by extraction
with methyl acetate or cyclohexane.  The graft copolymer obtained was
hydrolyzed, and the grafted polystyrene was isolated.

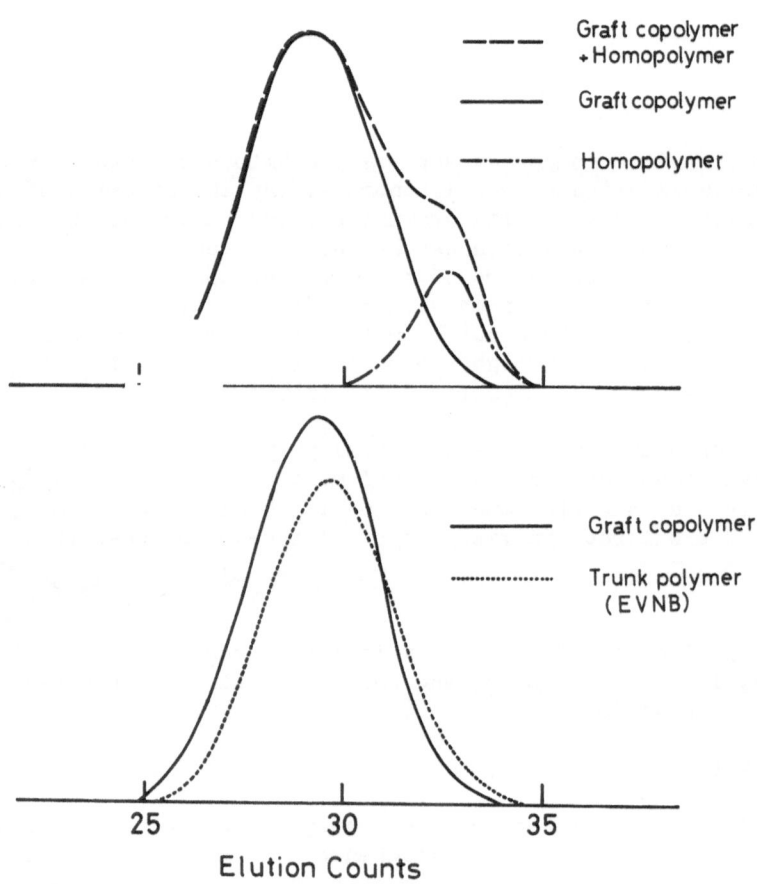

Figure 1.    GPC curves of graft copolymer, styrene homopolymer, a
             mixture of graft copolymer and homopolymer, and trunk
             polymer(EVNB) in the graft copolymerization of styrene
             onto ethylene-vinyl $p$-nitrobenzoate copolymer.

RESULTS AND DISCUSSION

GPC Curves

The change of GPC curve of ethylene-vinyl $p$-nitrobenzoate co-polymer (EVNB) by grafting polystyrene is shown in Figure 1. Two peaks appear for a mixture of graft copolymer and nongrafted poly-styrene. By extracting nongrafted polystyrene from the mixture with cyclohexane, the high-molecular weight peak is assigned to graft copolymer, and the low-molecular weight one is attributed to non-grafted polystyrene. The GPC curve for the trunk polymer EVNB expands to higher molecular weight by grafting polystyrene.

Reaction Conditions

Table 1 shows the results of graft copolymerization of styrene onto poly(vinyl $p$-nitrobenzoate) carried out at 60°C for various periods of time (4). As the reaction time was increased from 6 to 48 hr, the per cent grafting increased from 8% to 41%. No induction period was observed, so poly(vinyl $p$-nitrobenzoate) acts as a re-tarder in the polymerization of styrene. The graft efficiency was about 30% and almost independent of the reaction time. However, a slightly larger value of 38% was obtained at the shortest reaction time of 6 hr. The molecular weights of both grafted and nongrafted polystyrenes slightly increased with reaction time.

The number of monomer units of the trunk polymer per polystyrene branch decreased from 43 to 23 as the reaction proceeded. Namely, one polystyrene branch exists in every 23 monomer units on an average in the most highly branched graft copolymer obtained. The degree of polymerization of the trunk polymer was 970, so the number of poly-styrene branches per trunk polymer increased from 22 to 43.

These results indicate that the growing polystyrene radicals add to the nitro groups on the trunk polymer continually with reaction time, and highly branched graft copolymers can be prepared, when poly(vinyl $p$-nitrobenzoate) is used as a trunk polymer.

When the concentration of styrene in the polymerization mixtures was doubled while the other reaction conditions were kept almost the same (4), both the total conversion and the per cent grafting were increased compared to the results given in Table 1. The graft effi-ciency was as high as 95.4% for a reaction time of 3 hr, and then decreased with reaction time. The decrease in graft efficiency is due to the steric hindrance of the branches already formed. That is to say, the attack of the growing polymer radicals on the nitro groups of trunk polymer was prevented increasingly with the progress in grafting. For the same duration of reaction, more highly branched graft copolymer is obtained by doubling the concentration of styrene.

Table 1. Effect of Reaction Time on Graft Copolymerization of Styrene onto Poly(vinyl $p$-nitrobenzoate)

| Reaction time hr | Total conversion % | Per cent grafting % | Graft efficiency % | $\overline{M}_n \times 10^{-3}$ | | No. of nitro groups per branch | No. of branches per trunk polymer |
|---|---|---|---|---|---|---|---|
| | | | | Grafted PSt | Nongrafted PSt | | |
| 6 | 4.5 | 8.1 | 38.1 | — | — | — | — |
| 14 | 9.9 | 15.1 | 32.4 | 1.25 | 2.62 | 43 | 22 |
| 24 | 16.2 | 24.2 | 31.9 | 1.40 | 2.86 | 30 | 32 |
| 48 | 26.1 | 41.1 | 33.7 | 1.79 | 3.07 | 23 | 43 |

$[NO_2]$ = 9.01 x $10^{-2}$ mole/1.; $[St]$ = 7.81 x $10^{-1}$ mole/1.; $[AIBN]$ = 1.06 x $10^{-2}$ mole/1.; DMF, 50ml; 60°C.

Table 2. Effect of Azobisisobutyronitrile Concentration on Graft Copolymerization of Styrene onto Poly(vinyl p-nitrobenzoate)

| [AIBN] x 10³ mole/l. | Total conversion % | Per cent grafting % | Graft efficiency % | $\bar{M}_n$ x 10⁻³ | | No. of nitro groups per branch | No. of branches per trunk polymer |
|---|---|---|---|---|---|---|---|
| | | | | Grafted PSt | Nongrafted PSt | | |
| 4.85 | 4.7 | 40.7 | 92.0 | 7.85 | ---- | 104 | 9 |
| 6.79 | 5.6 | 37.8 | 71.2 | 5.83 | 2.70 | 84 | 12 |
| 9.69 | 6.8 | 31.3 | 48.5 | 2.70 | 3.51 | 47 | 21 |
| 14.5 | 7.5 | 16.8 | 23.8 | 1.96 | 3.65 | 42 | 23 |

$[NO_2]$ = 8.22 x 10⁻² mole/l.; $[St]$ = 1.43 mole/l.; DMF, 50ml; 60°C; 6 hr.

Table 3.   Effect of Amount of Trunk Polymer on Graft Copolymerization of Styrene onto Poly(vinyl $p$-Nitrobenzoate)

| $[NO_2] \times 10^2$ mole/1. | Total conversion % | Per cent grafting % | Graft efficiency % | $\overline{M}_n \times 10^{-3}$ Nongrafted PSt |
|---|---|---|---|---|
| 0 | 10.2 | ---- | --- | 4.19 |
| 2.25 | 8.3 | 15.3 | 9.9 | 2.71 |
| 4.51 | 6.5 | 11.5 | 18.8 | 2.24 |
| 6.67 | 4.9 | 9.2 | 30.0 | 1.83 |
| 9.01 | 4.5 | 8.1 | 38.1 | ---- |

$[St] = 7.81 \times 10^{-1}$ mole/1.; $[AIBN] = 1.06 \times 10^{-2}$ mole/1.; DMF, 50 ml; 60°C; 6 hr.

As the initiator concentration increased in the graft copolymerization of styrene onto poly(vinyl $p$-nitrobenzoate), the total
conversion increased, but the per cent grafting and the graft efficiency decreased (Table 2) (4).   The number of monomer units of
trunk polymer per polystyrene branch decreased from 104 to 42.   Accordingly, the number of branches per trunk polymer increases with
increasing initiator concentration, even if the graft efficiency
becomes lower.

When the concentration of poly(vinyl $p$-nitrobenzoate) as a
trunk polymer was increased, the graft efficiency increased linearly, but the total conversion and per cent grafting decreased,
and the molecular weight of nongrafted polystyrene decreased
(Table 3) (4).

The graft efficiency increases with increasing amount of the
trunk polymer because of the more frequent occurrence of the reaction between growing polystyrene radicals and the nitro groups
on the trunk polymer.   However, the total conversion, per cent
grafting and the molecular weight of nongrafted polystyrene
are reduced with the increase in the concentration of poly(vinyl
$p$-nitrobenzoate) due to the retardation of polymerization of styrene
by poly(vinyl $p$-nitrobenzoate) as mentioned above.

Trunk Polymer

We then used poly($p$-nitrophenyl acrylate) as a trunk polymer,
which has also pendant $p$-nitrophenyl groups connected to the polymer
backbone by ester bonds.

Table 4. Effect of Reaction Time on Graft Copolymerization of Styrene onto Poly(p-nitrophenyl Acrylate)

| Reaction time hr | Total conversion % | Per cent grafting % | Graft efficiency % | $\bar{M}_n \times 10^{-4}$ | | No. of nitro groups per branch | No. of branches per trunk polymer |
|---|---|---|---|---|---|---|---|
| | | | | Grafted PSt | Nongrafted PSt | | |
| 4 | 7.5 | 7.2 | 5.0 | --- | 2.14 | --- | --- |
| 8 | 14.3 | 12.0 | 4.6 | --- | 1.99 | --- | --- |
| 15 | 24.1 | 26.7 | 6.1 | 1.19 | 1.92 | 258 | 6.9 |
| 24 | 32.0 | 50.3 | 7.2 | 1.05 | 2.15 | 108 | 16.4 |

$[NO_2] = 6.47 \times 10^{-2}$ mole/l.; $[St] = 2.18$ mole/l.; $[AIBN] = 1.06 \times 10^{-2}$ mole/l.; DMF, 30ml; 60°C.

    As observed for poly(vinyl $p$-nitrobenzoate) as a trunk polymer,
the per cent grafting increased with reaction time, while the graft
efficiency was almost independent of reaction time.  However, the
graft efficiency is much smaller than that of about 30% observed for
poly(vinyl $p$-nitrobenzoate).  The molecular weights of both grafted
and nongrafted polystyrenes were ten times as high as those obtained
for poly(vinyl $p$-nitrobenzoate).  The number of monomer units of
the trunk polymer per branch was 108 even for 24 hr of reaction
time.  This value is quite larger than 23 obtained for poly(vinyl $p$-
nitrobenzoate) as a trunk polymer.  Poly($p$-nitrophenyl acrylate)
acts also as a retarder in the polymerization of styrene because no
induction period was observed.

    These results indicate that graft copolymer with a large
number of branches of low molecular weight can be prepared and the
graft efficiency is high, when poly(vinyl $p$-nitrobenzoate) is used
as a trunk polymer.  On the other hand, when poly($p$-nitrophenyl
acrylate) is used, graft copolymers obtained have a small number of
branches of high molecular weight and the graft efficiency is low.

    This difference is attributed to the polar effect of the ester
bond connecting the polymer backbone and pendant $p$-nitrophenyl
groups.  Therefore, several model compounds for trunk polymer with
pendant $p$-nitrophenyl groups were prepared, and the chain transfer
constants of polystyrene radicals to these model compounds were
determined by use of Kar's equation (5).

    As model compounds, we used isopropyl $p$-nitrobenzoate for
poly(vinyl $p$-nitrobenzoate), $p$-nitrophenyl isobutyrate for poly($p$-
nitrophenyl acrylate), $p$-nitrocumene for poly($p$-nitrostyrene), and
$p$-nitrophenyl isopropyl ether for poly($p$-nitrophenyl vinyl ether).
The structure and Hammett's $\sigma$ constants of these model compounds
and the chain transfer constants of polystyrene radicals to these
compounds are given in Table 5 (3).

    In Figure 2, the logarithmic chain transfer constant is plotted
against Hammett's $\sigma$ constant (3).  The chain transfer constant increases as the electron attracting property of the substituents
increases.  Although the experimental data are very limited, almost
linear relationship is observed, and the $\rho$ value of +1.4 is obtained
from the slope.

    Therefore, in these reaction systems, higher electron attracting
property of the substituent on benzene ring in $p$-position to the
nitro groups increases the reactivity of nitro groups to the growing polystyrene radicals resulting in high chain transfer constant.
The chain transfer constant to isopropyl $p$-nitrobenzoate as a model
compound for poly(vinyl $p$-nitrobenzoate) is the highest of the four
model compounds.  Accordingly, as the trunk polymer, poly(vinyl $p$-
nitrobenzoate) is expected to be most suitable for obtaining highly

Table 5.  Chain Transfer Constants of Polystyrene Radicals to
          Several Model Compounds for Trunk Polymers

| | $R^a$ | $C_x \times 10^2$ | Hammett's σ constant |
|---|---|---|---|
| p–Nitrophenyl isopropyl ether | $-O-C_3H_7$ | 3.8 | −0.45 |
| p–Nitrocumene | $-C_3H_7$ | 6.7 | −0.15 |
| p–Nitrophenyl isobutyrate | $-O-CO-C_3H_7$ | 9.7 | 0.16 |
| Isopropyl p–nitrobenzoate | $-CO-O-C_3H_7$ | 79.4 | 0.54 |

a   $NO_2$—⟨◯⟩—R

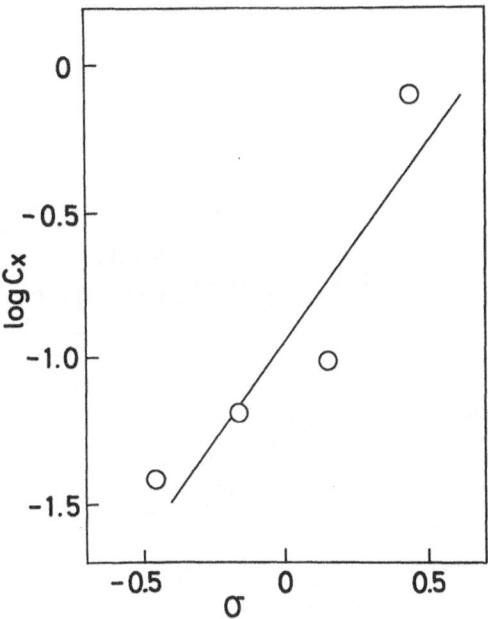

Figure 2.  Plot of $\log C_x$ vs. Hammett's σ constant for the model
           compounds of trunk polymers with pendant p–nitrophenyl
           groups.

branched graft copolymer of the four types of trunk polymers.  This
high reactivity of nitro groups on poly(vinyl $p$-nitrobenzoate) is
attributed to the high electron attracting property of the ester
groups of this trunk polymer compared to that of the ester bonds in
poly($p$-nitrophenyl acrylate).

Reaction Mechanism

     The reaction mechanism of low-molecular aromatic nitro compounds
as inhibitors or retarders with growing polymer radicals has been
extensively studied (1).  By summing up these results, the scheme
for the reaction of aromatic nitro groups on the trunk polymer with
growing polymer radicals is expected to be as given in Figure 3.

     In this reaction scheme, the growing polymer radicals initially
add to the oxygen atoms of the nitro groups and give intermediate
radicals(I).  Most of the radicals(I) decompose  through monomolecu-
lar mechanism to form nitroso groups.  Nitroso groups then combine
with one or two growing polymer radicals resulting in one or two
branches on a nitroso group (reaction A).  Part of the intermediate
radicals(I) terminate through a disproportionation (reaction B) or
recombination (reaction C) with growing polymer radicals yielding one
or two branches on a nitro group.  When the nitro group concentration
is high, the polymerization is reinitiated from the intermediate radicals and two branches are produced on a nitro group (reaction D).

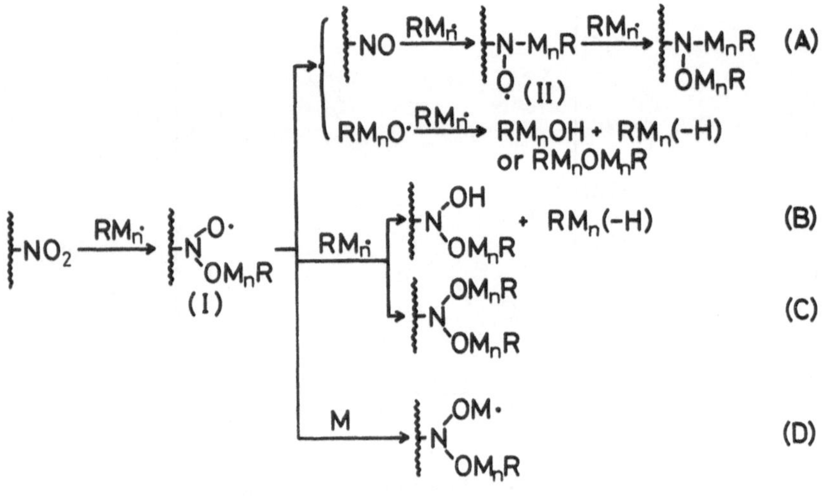

Figure 3.   Reaction scheme of aromatic nitro groups of trunk polymer
            with growing polymer radicals.

   In the reaction of $p$-nitrophenyl isopropyl ether with radicals
produced by the photo-induced decomposition of AIBN, the ESR spectrum
obtained is very similar to that observed for $p$-nitrosophenyl isopropyl ether under similar reaction conditions.  Accordingly, the
ESR spectra observed for these compounds correspond to that for the
nitroxide radicals(II) in the reaction scheme shown in Figure 3.

   Therefore, the graft copolymerization by chain transfer reaction
to $p$-nitrophenyl groups on the trunk polymer proceeds predominantly
through reaction (A), in which the intermediate is the nitroxide
radicals(II).  However, addition of the second polymer radicals to
the radicals(II) is hard to occur due to the steric hindrance as
will be discussed later.

## Relationship between Chain Transfer Constant, Monomer Conversion and Graft Frequency

   Taking the reaction mechanism mentioned above into consider-
ration, we derived the relationship between the chain transfer constant, extent of monomer conversion, and the number of branches in
the graft copolymerization by chain transfer mechanism (6).

   The number of nitro groups on the trunk polymer per branch, $n$
is expressed as

$$n = \frac{1}{a\left\{1 - (1 - \alpha)^{C_x}\right\}} \quad\quad (1)$$

where $a$ is the number of branches per nitro group reacted with
growing polymer radicals and assumes the value between 1 and 2,
$\alpha$ is the extent of monomer conversion, and $C_x$ is the chain transfer
constant of growing polymer radicals to the nitro groups on the
trunk polymer.

   In other words, $n$ means that a branch exists in every n monomer units on an average.  In equation (1), $a = 2$ means that all the
nitro groups reacted with growing polymer radicals have two branches,
and when $a$ equals     unity, all the nitro groups reacted have one
branch.  If the value of $a$ is known, the value of $n$ may be predicted
from the chain transfer constant $C_x$ and the extent of monomer conversion $\alpha$ for a given monomer-nitro group containing trunk polymer
system.

   However, it is difficult to determine the value of $a$ from
experimental data.  As seen from Figure 4, even if two branches are
formed on a nitro group of trunk polymer, hydrolysis of the ester
bond yields one grafted polymer.

Then, $a'$ is defined as the ratio of the nitro groups having branches to those reacted with growing polymer radicals as follows:

$$a' = \frac{P_{G'}}{1 - (1 - \alpha)^{C_x}} \qquad (2)$$

where $P_{G'}$ is the number of nitro groups which have one or two branches divided by the initial number of pendant nitro groups. That is to say, $P_{G'}$ is the fraction of nitro groups having branches.

On the contrary to the value of $a$, the value of $a'$ can be determined experimentally by separating branches from the trunk polymer by cleaving ester groups in the case of poly(vinyl $p$-nitrobenzoate) and poly($p$-nitrophenyl acrylate).

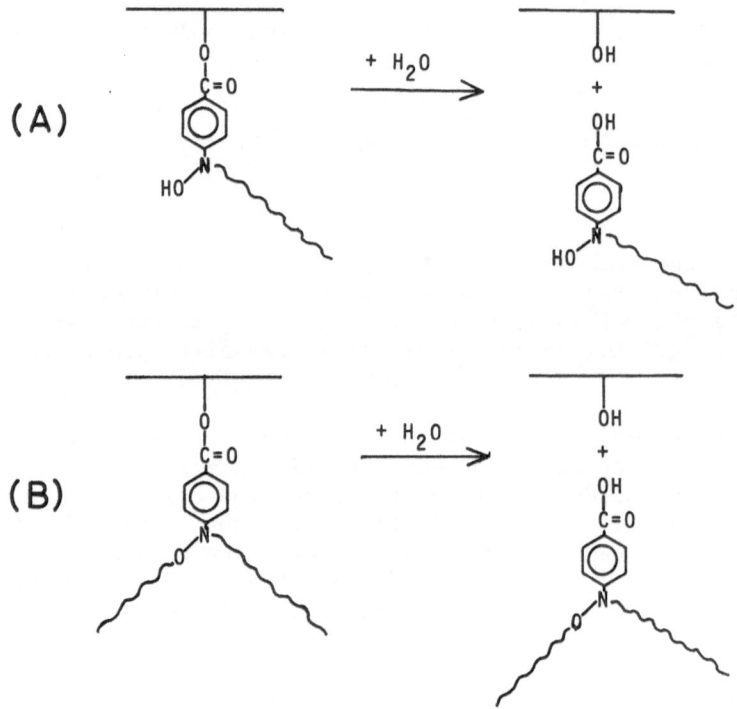

Figure 4. Hydrolysis of polystyrene-grafted poly(vinyl $p$-nitrobenzoate) with (A) one or (B) two branches on a nitro group of trunk polymer.

Table 6. Values of $P_{G'}$, and $\alpha'$ Obtained for the Graft Copolymerization of Styrene onto Poly(vinyl p-Nitrobenzoate)

| [NO$_2$] x 10$^2$ mole/l. | [St] x 10 mole/l. | [AIBN] x 10$^2$ mole/l. | Time hr | $\alpha$ x 10$^2$ | $P_{G'}$ x 10$^3$ | $\frac{c}{[1-(1-\alpha)^x]}$ x 10$^3$ | $\alpha'$ |
|---|---|---|---|---|---|---|---|
| 9.01 | 7.81 | 1.06 | 14 | 9.9 | 23.3 | 33.3 | 0.700 |
| | | | 24 | 15.2 | 33.3 | 56.4 | 0.594 |
| | | | 48 | 26.1 | 43.5 | 94.2 | 0.462 |
| 8.22 | 14.3 | 0.969 | 6 | 6.8 | 21.3 | 22.7 | 0.938 |
| | | | 15 | 15.4 | 25.6 | 53.2 | 0.481 |
| | | | 24 | 22.3 | 37.0 | 79.0 | 0.468 |
| | | 0.485 | 6 | 4.7 | 9.61 | 15.6 | 0.616 |
| | | 0.679 | 6 | 5.6 | 11.9 | 18.4 | 0.647 |
| | | 0.969 | 6 | 6.8 | 21.3 | 22.7 | 0.938 |
| | | 1.45 | 6 | 7.5 | 23.8 | 25.2 | 0.944 |

The values of $a'$ were calculated for the results on the graft copolymerization of styrene onto poly(vinyl $p$-nitrobenzoate) (Table 6) (6). The values of $a'$ decrease with reaction time and become smaller than 0.5 for longer reaction times. Moreover, $a'$ depends on the concentration of initiator. At higher initiator concentrations $a'$ is very close to unity, whereas at low concentrations $a'$ becomes considerably smaller.

The decrease in the value of $a'$ with reaction time, namely with increase in per cent grafting, is due to the steric hindrance of branches already formed on the trunk polymer. Attack of the growing polymer radicals on the nitro groups on the trunk polymer is prevented increasingly by the branches already formed, and the apparent chain transfer constant to the nitro groups on the trunk polymer is decreased.

Steric Hindrance

The values of chain transfer constants of polystyrene radicals to the pendant nitro groups of the trunk polymers are less than half of those to their model compounds (3,4). This difference is attributed to the difficulty of the bulky polystyrene growing radicals to diffuse into the random coil of trunk polymer and to attack the nitro groups inside the random coil due to the steric hindrance. As already mentioned, the steric hindrance plays an important role in attack of growing polymer radicals to the pendant nitro groups on the trunk polymer. Therefore, when the distribution of pendant nitro groups on the trunk polymer is less frequent, the nitro groups on the trunk polymer will be used more efficiently in the graft copolymerization by chain transfer reaction.

We then carried out graft copolymerization of styrene onto ethylene-vinyl $p$-nitrobenzoate copolymer, whose content of vinyl $p$-nitrobenzoate was 28.6 mole% (Table 7).

The chain transfer constant of polystyrene radicals to ethylene-vinyl $p$-nitrobenzoate copolymer was compared with those to poly(vinyl $p$-nitrobenzoate) and their model compound, isopropyl $p$-nitrobenzoate. As expected, the value of chain transfer constant to the copolymer is larger than that to poly(vinyl $p$-nitrobenzoate) and smaller than that to the model compound isopropyl $p$-nitrobenzoate obtained according to the method of Mayo et al. (7) (Table 8).

As seen from Table 7, the graft efficiency was higher than that of about 30% for poly(vinyl $p$-nitrobenzoate). The number of nitro groups per polystyrene branch was 28 for 24 hr and 20 for 48 hr of the reaction time. These values are smaller than 30 and 23 observed for poly(vinyl $p$-nitrobenzoate). Therefore, the nitro groups on ethylene-vinyl $p$-nitrobenzoate copolymer are used more efficiently than those on poly(vinyl $p$-nitrobenzoate). The values of $a'$ were

Table 7. Effect of Reaction Time on Graft Copolymerization onto Ethylene-Vinyl p-Nitrobenzoate Copolymer

| Reaction time | Total conversion | Per cent grafting | Graft efficiency | $\overline{M}_n \times 10^{-3}$ | | No. of nitro groups per branch | $a'$ |
|---|---|---|---|---|---|---|---|
| | | | | Grafted PSt | Nongrafted PSt | | |
| hr | % | % | % | | | | |
| 6 | 5.2 | 8.9 | 52.6 | ---- | 1.75 | -- | ----- |
| 15 | 7.8 | 16.7 | 47.4 | 1.95 | 2.38 | 44 | 0.609 |
| 24 | 11.4 | 24.9 | 48.1 | 1.83 | 2.80 | 28 | 0.649 |
| 48 | 17.4 | 34.9 | 44.3 | 1.86 | 2.82 | 20 | 0.584 |

[NO$_2$] = 6.89 x 10$^{-2}$mole/l.; [St] = 7.92 x 10$^{-1}$mole/l.; [AIBN] = 1.11 x 10$^{-2}$mole/l.; DMF, 50ml; 60°C.

Table 8. Chain Transfer Constants of Polystyrene Radicals
to Poly(vinyl $p$-Nitrobenzoate), Ethylene-Vinyl $p$-
Nitrobenzoate Copolymer and Isopropyl $p$-Nitro-
benzoate

|  | $C_x$ |
|---|---|
| Poly(vinyl $p$-nitrobenzoate) | 0.357 |
| Ethylene-vinyl $p$-nitrobenzoate copolymer | 0.468 |
| Isopropyl $p$-nitrobenzoate | 0.736 |

also higher than those obtained for poly(vinyl $p$-nitrobenzoate)
under similar reaction conditions.

CONCLUSIONS

   A graft copolymer can be prepared through the chain transfer
reaction of growing polymer radicals to the aromatic nitro groups,
when a vinyl monomer is polymerized with a radical initiator in the
presence of a polymer with pendant aromatic nitro groups.  Especially when poly(vinyl $p$-nitrobenzoate) is used as a trunk polymer,
highly branched graft copolymers are obtained due to the high electron attracting property of the ester bonds connecting $p$-nitro-
phenyl groups and the polymer backbone.

   Steric hindrance plays an important role in this method for
preparing graft copolymers, as the attack of growing polymer radicals to the pendant nitro groups on the trunk polymer is prevented
increasingly with the progress in grafting.  Therefore, the less
frequently the pendant nitro groups are distributed on the trunk
polymer, the more efficiently the nitro groups are utilized in the
graft copolymerization.

References

1.  G. Goldfinger, W. Yee, and R. D. Gilbert, in "Encyclopedia of
    Polymer Science and Technology," Vol. 7, H. F. Mark, N. G.
    Gaylord, and N. M. Bikales, ed., Interscience, New York
    (1967), p.644.
2.  S. Nakamura, H. Sato, and K. Matsuzaki, J. Polym. Sci. B 11:221
    (1973).
3.  S. Nakamura, H. Kasatani, and K. Matsuzaki, J. Appl. Polym. Sci.
    in press.
4.  S. Nakamura, H. Sato, and K. Matsuzaki, J. Appl. Polym. Sci. 20:
    1501 (1976).

5.  I. Kar, B. M. Mandal, and S. R. Palit, Makromol. Chem. 127:195
    (1969).
6.  S. Nakamura, M. Yamada, and K. Matsuzaki, J. Appl. Polym. Sci.
    22:2011 (1978).
7.  R. F. Mayo, R. A. Gregg, and M. S. Matheson, J. Am. Chem. Soc.
    73:1691 (1951).

# STYRENE COPOLYMERIZATION WITH RUBBER. I. STUDIES OF POLYSTYRENE GRAFTING TO RUBBER: MOLECULAR WEIGHT CHARACTERISTICS OF HOMO-POLYSTYRENE MATRIX

V. D. Yenalyev, V. I. Melnichenko, N. A. Noskova,
O. P. Bovkunenko, C. I. Yegorova, N. G. Podosenova
and V. P. Budtov

Donetsk State University
Donetsk, 340055, USSR

Styrene copolymerization with rubber initiated by a mixture of peroxides with different thermal stabilities is widely used for preparing HIPS (high impact polystyrene) in industry. Such initiation allows regulation of the process rate at all stages and wide variation of the physical and chemical properties of the material. Full knowledge of qualitative regularities in molecular weight as well as structural parameters of formation of polymer compositions provides grounds for theoretical consideration of the problem of optimal steering reactions of polymer synthesis. However, these data for styrene copolymerization with elastomers in the presence of a peroxide mixture are, for all practical purposes, not available in the literature. This constitutes the subject of this research work.

Styrene copolymerization with polybutadiene (ISR "inten-55 NFA") was carried out by the bulk method while agitated mechanically. The rate of agitation was retained constant up to polymerization conversion of 20-25%. The polymerizations were carried out in sealed glass ampoules. Styrene of 99.8% purity was distilled in an inert gas atmosphere three times. Rubber was dissolved in styrene by mixing for 3 hours at 60°C. Initiators were then added into the reaction system in order to achieve a copolymerization process in two temperature regimes. Benzoyl peroxide (BP) and t-butyl perbenzoate (TBPB), having different intervals of "working temperatures," were chosen as initiators. The choice of concentrations of peroxides and temperatures at both stages as well as evaluating the experimental data was achieved by the method of statistic experiment planning (1). This allows us to cover widest areas of research and to obtain quantitative dependences using a minimal number of experiments.

Thus the following experimental process parameters were chosen:

$X_1$ - benzoyl peroxide concentration
$X_2$ - t-butyl perbenzoate concentration
$X_3$ - the first stage polymerization temperature ($t_1$)
$X_4$ - the second stage polymerization temperature ($t_2$)
$X_5$ - final first-stage monomer conversion.

The intervals of variable variation are given in Table I.

The degree of grafting polystyrene onto polybutadiene was determined by the method of selective dissolving (2); the molecular weight (MW) and molecular weight distribution (MWD) of homopolystyrene was measured using gel permeation chromotography (GPC)(3).

From the experimental data, we obtained the regression equations connecting the molecular weight characteristics of polystyrene matrix, the degree of grafting of polystyrene onto rubber, and the equilibrium degree of gel swelling (which define the physical and mechanical properties of HIPS) with parameters from the polymerization initiation.

For Mw, Mz, Mw/Mn, Mz/Mw the equations are as follows:

$$
\begin{aligned}
Mw \cdot 10^{-5} = {} & 2.62 - 1.15X_1 - 0.35X_2 - 0.20X_3 - 0.62X_4 + \\
& 0.16X_5 + 0.15X_1X_2 + 0.14X_1X_3 + 0.42X_1X_4 - \\
& 0.12X_1X_5 + 0.14X_2X_3 - 0.19X_3X_5 + \\
& 0.32X_1^2 + 0.08X_2^2 + 0.20X_3^2;
\end{aligned}
\tag{1}
$$

$$
\begin{aligned}
Mz \cdot 10^{-5} = {} & 5.62 - 2.08X_1 - 0.61X_2 - 0.39X_3 - 1.16X_4 + \\
& 0.16X_5 + 0.09X_1X_2 + 0.51X_1X_4 + 0.20X_1X_5 + \\
& 0.43X_2X_3 + 0.20X_3X_4 - 0.42X_3X_5 + 0.55X_1^2 + \\
& 0.26X_2^2 + 0.25X_3^2 - 0.15X_4^2;
\end{aligned}
\tag{2}
$$

$$
\begin{aligned}
\frac{Mw}{Mn} = {} & 2.50 - 0.07X_1 + 0.11X_2 - 0.26X_3 + 0.07X_4 - \\
& 0.13X_5 - 0.28X_1X_2 - 0.10X_1X_3 + 0.23X_1X_5 + \\
& 0.09X_2X_3 - 0.09X_3X_4 + 0.13X_1^2 + 0.33X_2^2 + 0.19X_4^2 + \\
& 0.55X_5^2;
\end{aligned}
\tag{3}
$$

$$\frac{Mz}{Mw} = 2.12 + 0.06X_1 + 0.03X_2 - 0.04X_1X_2 - 0.06X_1X_3 -$$

$$0.07X_1X_4 + 0.11X_1X_5 + 0.03X_2X_3 + 0.04X_2X_4 +$$

$$0.03X_3X_4 + 0.04X_2^2 - 0.05X_3^2 - 0.04X_4^2 \tag{4}$$

The above equations provide the grounds to calculate the influence of one of the factors on the polymer properties in the field of variables variation, fixing the rest at a definite level, and to investigate more deeply the essence of the obtained regularities, and the mechanism of the polymerization process. The equations and the calculated curves reflect some already known dependences of molecular weight characteristics on process parameters. Thus, the increase of BP concentration and the temperature $t_1$, results in the lowering of homopolystyrene MW, whereas the ratio Mw/Mn varies insufficiently.

A number of peculiarities in the investigated process were found which must be considered in detail. Figure 1 shows dependences of Mw, Mw/Mn, Mz and Mz/Mw on first stage monomer conversion and TBPB concentration, as well as experimental points for the corresponding levels. The growth of conversion up to 40-50% in isothermal conditions is known (4) to influence insufficiently the MW of the homopolystyrene being formed. That is why the observed changes of MWD in the product (Figure 1, a, b) are defined by the conditions of the second stage polymerization. Decrease of Mw/Mn and a practically constant value of Mw at the first stage up to conversion ∼ 30-40% shows that the quality of molecular homopolystyrene fractions received at the second stage is being diminished due to the influence of the gel effect.

But the increase of the final first stage conversion up to 50-65% results in the increase of the ratio TBPB:monomer, which results in the lowering of polystyrene Mn at the second stage of polymerization. It promotes in its turn a large increase of the

| Levels | BP mole/l · $10^2$ | TBPB mole/l · $10^2$ | $t_1$ °C | $t_2$ °C | S, % |
|--------|------|------|------|------|------|
| 0 | 0.6 | 0.6 | 80 | 115 | 45 |
| + 1 | 1.0 | 0.9 | 90 | 125 | 55 |
| - 1 | 0.2 | 0.3 | 70 | 105 | 35 |
| + 1.61 | 1.244 | 1.083 | 96.1 | 121.1 | 61.1 |
| - 1.61 | 0 | 0.117 | 63.9 | 98.9 | 28.9 |

Table I.  The intervals of variables variation.

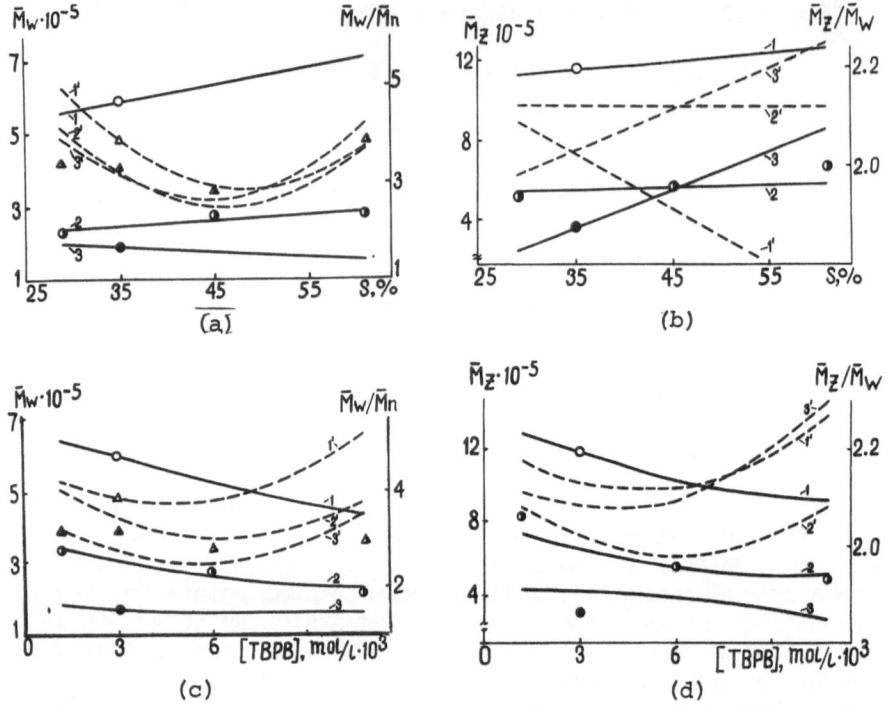

Figure 1.   Dependence of Mw, Mw/Mn, Mz, Mz/Mw on:   (a,b) final
            first stage monomer conversion;   (c,d,) TBPB concen-
            tration.   The curves I (◯), 2 (◑), 3 (●) for Mw and Mz,
            and 1' (△), 2' (▲), 3' (▲) for Mw/Mn and correspond
            to the conditions of low, middle and high levels of
            variables.   Designation of experimental points is
            given in parentheses.


ratio Mw/Mn at different BP concentrations and values of $t_1$.

        The height of the first stage polymerization influences the
values of Mz and Mz/Mw in a complicated way.   Figure 1-b shows
that the polymerization conditions change not only the quantity
of high molecular weight polymer fractions but also the direction
of the change with monomer conversion on Mz and Mz/Mw.   The BP
concentration, temperature and polymerization duration at the first
stage define the extent of thermal and induced TBPB decomposition
or, on the other hand, its content in the system after the first
stage of polymerization is over.   The given conditions and temperature at the second stage will define this initiation concentration at the final polymerization stage (after 90%), which in its
turn will influence the quantity of high molecular weight fractions
of polystyrene matrix being formed at this period.

At low TBPB concentrations, homopolystyrene formed at the second stage has much higher MW than at the first. In this case there is no dependence on BP concentration and temperature $t_1$ which results in high polydispersity of the polystyrene matrix. The increase in TBPB concentration results in lowering polystyrene MW and suppression of the gel effect as a result of which polystyrene MW is almost equal to the MW of the one being formed at both stages. This causes some reduction of the Mw/Mn and Mz/Mw values (see Figure 1-c,d). The further increase of TBPB concentration in the system promotes formation of low molecular fractions at the second stage, which is verified by the large increase of homopolystyrene dispersity in the HIPS. This is how the extreme character of the dependences of Mw/Mn and Mz/Mw on the polymerization height at the first stage and on TBPB is explained.

Variation of the parameters given in Table I defines not only molecular weight characteristics of the polystyrene matrix but also composition and structure of the graft copolymer. Regression equations of the following type were obtained for degree of grafting, Gs, and equilibrium degree of gel swelling, Q:

$$Gs = 314 + 16.8X_2 + 5.7X_3 + 36.8X_4 - 15.0X_1X_2 -$$

$$17.3X_1X_3 - 6.5X_1X_5 + 13.1X_2X_4 - 10.5X_3X_4 +$$

$$15.5X_3X_5 + 3.4X_4X_5 - 14.1X_1^2 - 10.1X_3^2 +$$

$$10.2X_4^2 + 10.2X_5^2; \tag{5}$$

$$Q = 15.2 + X_1 - 1.1X_2 - 0.6X_3 - 2.3X_4 - 1.4X_2X_4 +$$

$$0.2X_1^2 - 0.3X_2^2 + 0.6X_3^2 + 0.6X_4^2 \tag{6}$$

It can be seen from equation (5) that the dependence of grafting degree on all variables has a complicated character. It can be illustrated, for example, by its dependence on BP concentration and the first stage temperature for particular cases (Figure 2).

The increase of BP concentration causes some growth of grafting degree at low copolymerization temperatures, as has been mentioned in other papers (5,6). However, Figure 1-a shows that further increase of BP concentration stops the growth of grafting and even diminishes it. The minimum location in the curves depends on the temperature process; as it is increasing, the minimum is shifted to the side of low BP concentration. The dependence of the graft degree on polymerization temperature is of analogous character at the first stage. The essence of the marked regular-

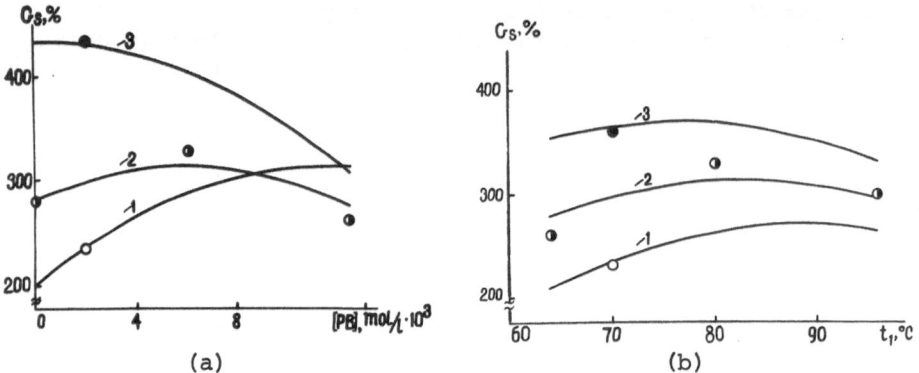

Figure 2.   Relationship of grafting degree and BP (a) and the
            first stage temperature (b).  The curves 1(O), 2(◐),
            3(●) correspond to the conditions of the low, middle
            and high levels of variables.  Designation of experi-
            mental points is given in parentheses.

ities is likely to be in the structure of copolymer macromolecules
being formed.

    The increase of both BP concentration and the temperature
promotes the growth of the number of polystyrene branches being
grafted onto the rubber with the simultaneous lowering of their
length at the first stage.  Such a growth of the number of grafted
branches changes the polybutadiene structure in the reaction
medium in a considerable way.  As a consequence of steric difficulties, the possibility of new active centers being formed on
the elastomer macromolecules is lowered.  As a result, the graft
reaction is considerably slowed with further polymerization.  In
these cases the lowering of gel crosslinking at the high stages
is noted, which is verified by the curves of equilibrium degree
of gel swelling BP concentration (Figure 3-a).

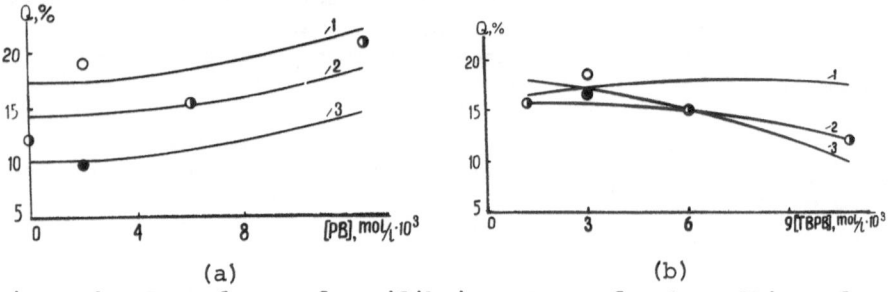

Figure 3.   Dependence of equilibrium stage of gel swelling of
            HIPS on BP (a) concentration and TBPB (b) concentration:
            the curves 1(O), 2(◐), 3(●) correspond to the conditions
            of low, middle and high levels of variables.  Designa-
            tion of experimental points is given in parentheses.

   The equilibrium swelling degrees vs. TBPB concentration
curves (Figure 3-b) agree with the known view mechanisms of free
radical crosslinking of elastomers (7,8).

   The kinetics of formation of graft copolymer and networks
at the second stage is defined by the TBPB concentration and
temperature. The increase of each parameter, due to the corresponding growth of the grafting degree and the density of polymer
network gel in HIPS, results in a maximum value of Gs. This depends both on copolymerization rate at the second stage and the
emulsion structure which had been formed at this moment. The dependence of rate of change of graft copolymer formation in the
course of copolymerization of styrene with polybutadiene on conditions at the second stage is represented in Figure 4. This
figure also shows that an appreciable quantity of graft copolymer is formed at the high polymerization stages. The increase in

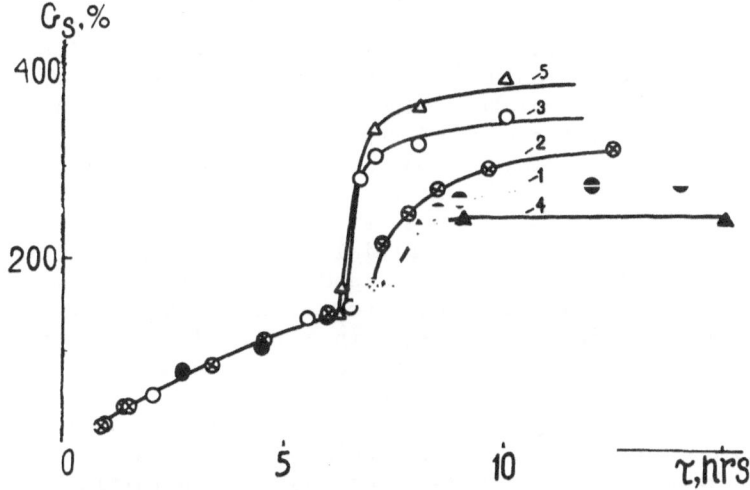

Figure 4.  The change of grafting degree in the course of styrene
           copolymerization with rubber:
           1 - TBPB = $0.117.10^{-2}$ mol/l, $t_2$ = $115^{\circ}$C.
           2 - TBPB = $0.6.10^{-2}$ mol/l, $t_2$ = $115^{\circ}$C.
           3 - TBPB = $1.244.10^{-2}$ mol/l, $t_2$ = $115^{\circ}$C.
           4 - TBPB = $0.6.10^{-2}$ mol/l, $t_2$ = $98.9^{\circ}$C.
           5 - TBPB = $0.6.10^{-2}$ mol/l, $t_2$ = $131.1^{\circ}$C.
           The values of the other variables correspond to the
           middle level, given in Table I.

temperature from 100°C to 130°C at the second stage results in the
increase of grafting degree in HIPS from 260 to 400%, i.e., approximately 1.5 times as much at the other equal conditions (see curves
4 and 5, Figure 4). Such an increase of the degree of grafting
may be explained in the following way: in accordance with the
data obtained by us earlier for isothermal styrene copolymerization
with polybutadiene (9), the rates of homopolystyrene formation
and graft copolymer become practically equal at the high stages
of the process, although at the initial stages, homopolymer ac-
cumulation prevails over the formation of graft copolymer in a
considerable way.

     The high grafting rate as compared to styrene homopolymeri-
zation can be explained as follows: the activation energy of
formation of active centers on polybutadiene macromolecules is
considerably higher than the activation energy of homopolystyrene
chain propagation (according to our data, it is approximately 12
kcal/mol); the increase in temperature will increase the rate of
graft copolymer formation to a larger degree than styrene homo-
polymerization. In addition, the peculiarities of the process
of graft copolymer formation at the high stages are affected by
the non-continuity of the rubber phase, its inner structure, and
can be explained by the following: based on the general rate
values of polymerization at the high stages (60-80% of styrene
conversion) and rate constants of elementary reactions, the possible number of free radicals found in 1 $cm^3$ of reaction mass volume
equalling $10^{13}$ was calculated. Then, based on the general volume
of the rubber phase, the size of its particles, the occlusion size
and the initial polybutadiene content, we calculated the number of
polystyrene occlusions in the unit of volume of the rubber phase.
For example, when the rubber phase volume was 20%, the size of
rubber particles was $1\mu$ , the initial polybutadiene concentration
7% and the size of occlusions $0.1\mu$ and $0.2\mu$ , their quantity in
1 $cm^3$ makes up 2.5 x $10^{14}$. Thus, one can see that inside the rubber phase there is only one radical in 3-30 occlusions. Since
diffusion of a free radical beyond the occlusion boundary is made
difficult as a result of high polybutadiene concentration in mem-
branes between neighboring occlusions, the probability of the
square termination of radicals is negligible. Hence, the radical
which was formed in one or several occlusions can initiate the
growth of chains many times, practically each act(ion) of repeated
initiation yielding active centers on polybutadiene macromolecules.
This in its turn will increase the rate of grafting and promote
the formation of networks at the high polymerization stages.

     Thus, it is seen that the choice of synthesis conditions of
HIPS with desired physical and chemical properties should be made
based on the mutual influence of all factors which define the process of styrene copolymerization with polybutadiene. The solution
of this problem becomes easier when regression equations are used.

REFERENCES

1.  V. V. Nalimov, N. A. Tchernova, Statistic Methods of Planning Extremal Experiments, Nauka, Moscow (1965).
2.  Y. Mori, V. Minoura, M. Imoto, Macromol. Chem., 25, 1 (1958).
3.  V. P. Budtov, N. G. Podosenova, E. I. Yegorova, Vysokomol. Soed., A XIX, 2160 (1977).
4.  V. A. Melnichenko, Candidate Thesis, Donetsk (1976).
5.  P. W. Allen, F. M. Merett, J. Polym. Sci., 22, 193 (1956).
6.  J. L. Locatelli, G. Riess, Angew. Makromol. Chem., 23, 161 (1973).
7.  L. D. Loan, Rubber Chem. and Technol., 40, 149 (1967).
8.  D. J. Stein, G. Fahrbach, H. Adler, Angew. Makromol. Chem., 38, 67 (1974).
9.  V. D. Yenalyev, N. A. Noskova, V. I. Melnichenko, ACS Polym. Prepr., 37, 645 (1977).

STYRENE COPOLYMERIZATION WITH RUBBER. II. RELATIONSHIP BETWEEN

POLYMER MORPHOLOGY AND SYNTHESIS CONDITIONS.

V. D. Yenalyev, N. A. Noskova, V. I. Melnichenko,
Y. N. Zhuravel and V. M. Bulatova

Donetsk State University
Donetsk 340055, USSR

Recently published papers (1-4) show that the morphology
(the size of dispersed phase particles, its volume fraction and
the structure of the polymer intermediate layers) of composite
polymer materials definitely influences their mechanical resistance.
Our research on the microstructure of high impact polystyrene
(HIPS) prepared by thermally initiated copolymerization of styrene
with rubber (5) has shown that the microstructure can be changed
up to high stages of polymerization. The morphological parameters
of HIPS, prepared by styrene copolymerization with polybutadiene
in the presence of a mixture of initiators, were studied side by
side with the effects of molecular weight characteristics of the
polystyrene matrix and grafted copolymer.

The synthesis conditions of the polymer according to the
statistical experimental plan and the values of the parameters
variables of the process were given in the previous chapter. The
HIPS microstructure was studied by electron microscopy (5). The
results are given in Table I. The values of variables at the
upper and lower levels are also shown.

The regression equations resulting from the experiments are
as follows:

$$\overline{V}_f = 32.9 + 2.2X_1 + X_1X_2 - 5.2X_1X_3 + X_4X_5 - 3.6X_1^2 +$$
$$1.4X_2^2 - 5.8X_3^2 + X_5^2; \tag{1}$$

$$\overline{C} = 1.7 - 0.20X_3 - 0.28X_1X_3 + 0.10X_2X_3 - 0.42X_1^2 -$$
$$0.45X_3^2 \tag{2}$$

| $\dfrac{BP}{TBPB}$, mol/l | | $\dfrac{0.2 \cdot 10^{-2}}{0.3 \cdot 10^{-2}}$ | | $\dfrac{1.0 \cdot 10^{-2}}{0.3 \cdot 10^{-2}}$ | | $\dfrac{1.0 \cdot 10^{-2}}{0.9 \cdot 10^{-2}}$ | | $\dfrac{0.2 \cdot 10^{-2}}{0.9 \cdot 10^{-2}}$ | |
|---|---|---|---|---|---|---|---|---|---|
| $t_1$,°C | S, % | $t_2$,°C | $\overline{Vf}$, %/$\overline{C}$ | $t_2$,°C | $\overline{Vf}$, %/$\overline{C}$ | $t_2$,°C | $\overline{Vf}$, %/$\overline{C}$ | $t_2$,°C | $\overline{V}$, %/$\overline{C}$ |
| 70 | 35 | 105 | 21.5/0.73 | 125 | 31.3/1.28 | 105 | 33.0/1.06 | 125 | 19.9/0.56 |
| | 55 | 125 | 23.4/0.81 | 105 | 30.4/1.22 | 125 | 35.4/1.14 | 105 | 15.7/0.51 |
| 90 | 35 | 125 | 28.1/0.73 | 105 | 20.3/0.46 | 125 | 23.4/0.39 | 105 | 32.3/1.23 |
| | 55 | 105 | 28.9/0.81 | 125 | 19.9/0.43 | 105 | 22.8/0.39 | 125 | 31.1/1.25 |

Table I.   Dependence of Rubber Phase Volume ($\overline{Vf}$) and Dispersed Particles Size ($\overline{C}$) on Conditions of Initiated Styrene Copolymerization with Rubber.

The rest coefficients of the twin relationship and the square terms are omitted because of their not being essential (according to Student criterion). Dependencies of rubber phase fraction volume and the mean size of dispersed particles on every researched parameter of the process were calculated in order to fix the rest of the parameters at low, middle and upper levels (presented in Figure 1). The experimentally found values $\overline{V}f$ and $\overline{C}$ for the corresponding conditions are marked with points for comparison here as well.

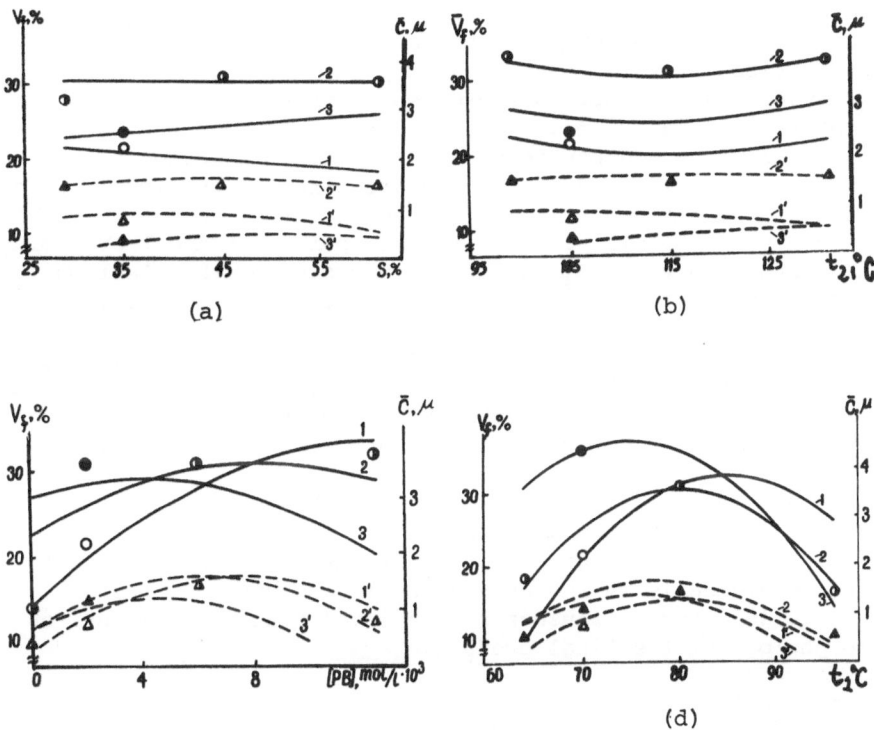

(a)     (b)

(d)

Figure 1. The calculated dependences of rubber phase volume and the mean size dispersed particles on: (a) final conversion at the first stage; (b) temperature at the second stage; (c) benzoyl peroxide (BP) concentration; (d) temperature at the first stage. The curves 1 (O), 2 (◑), 3 (●) for $V_f$ and 1' (△), 2' (▲), 3' (▲), for C correspond to the conditions of low, middle and upper levels of variables. Designation of experimental points is given in parentheses.

Considering the influence of the final first stage conversion and polymerization temperature at the second stage (Fig. 1, a, b) on $\overline{VF}$ and $\overline{C}$, i.e., parameters defining the copolymerization reaction at the middle and high stages, it is seen that their variation in the chosen intervals doesn't bring any essential changes in morphology of the polymer. At first appearance this may seem to contradict the conclusion made for the styrene thermal copolymerization with rubber (5). However, initiated polymerization at low temperatures (t$\leqslant$125$^{\circ}$C) results in reaction mass viscosities 2-8 times higher than thermal polymerization. Both high viscosity and low temperature appreciably decrease the diffusion coefficient and limit the possibility of mass exchange between rubber and polystyrene phases in emulsion polymerization. The polymerization rate at the second stage (Fig. 2) is rather high even at the smallest initiator concentrations and minimal temperature $t_2$ (Fig. 2, curve 7). This explains why the external sizes of the dispersed particles and the volume fraction of the rubber phase in all cases considered are fixed at the level which was achieved at the beginning of the second stage completion. The independence of the size of rubber phase particles and their volume on the conversion value $S_1$ is also illustrated by the electron micrographs shown in Fig. 3. This indicates that equilibrium is already being disturbed at the first stage because of poor mass exchange, and the structure is stabilized before the first stage if over. At the same time, when the thermal initiation takes place, a decrease of approximately 25% in $\overline{Vf}$ is observed in the interval of monomer conversion between 35 and 55%.

It now becomes necessary to discuss the dependences of $\overline{Vf}$ and $\overline{C}$ on BP concentration and temperature at the first stage. The fact that the emulsion structure at the second stage is not changed considerably in the chosen conditions gives us the grounds to judge the influence of the investigated parameters on the morphology of the final product.

Figure 1 shows that the plots of structural parameters of HIPS vs. BP concentration and the value of $t_1$ go through maxima, the position of which shifts to the side of diminishing variables values with the rise of the level. This may be explained by taking into consideration that the main role in the formation of HIPS morphology is played by the quantity and structure of the graft copolymer. The increase of BP concentration and temperature $t_1$ results in the growth of grafting rate at the beginning of polymerization when the system represents mostly the continuous phase. It is known (8) that about 80% of rubber is transformed into the graft copolymer by 30% of monomer conversion. The calculations of polybutadiene distribution on the volume of dispersed particles made on the basis of the value of the interface surface, the probable size of macromolecular polybutadiene chains and its total

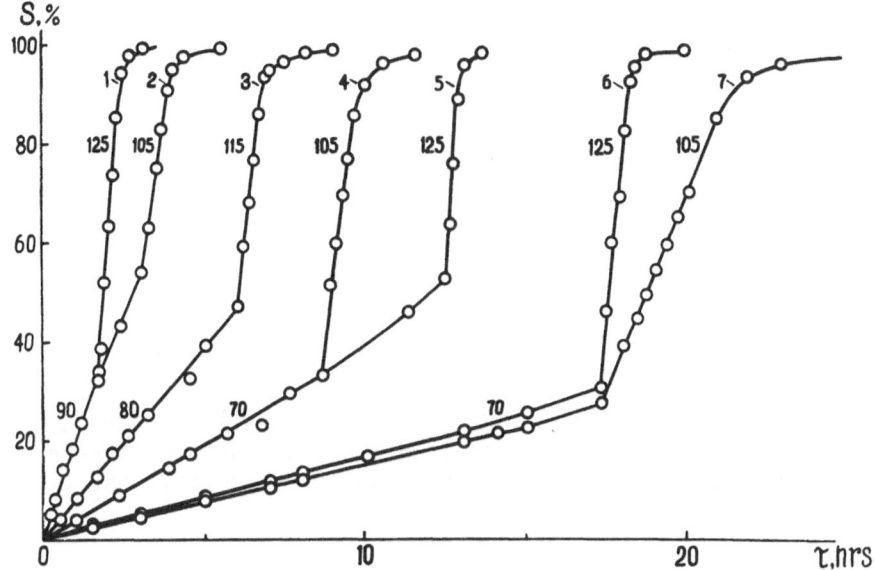

Figure 2.   Dependence of monomer conversion on polymerization
            time when the benzoyl peroxide and TBPB are relatively
            in moles/l:   1,2) $1.0 \cdot 10^{-2}$ and $0.9 \cdot 10^{-2}$;
            3) $0.6 \cdot 10^{-2}$ and $0.6 \cdot 10^{-2}$;    4,5) $1.0 \cdot 10^{-2}$ and
            $0.3 \cdot 10^{-2}$;    6) $0.2 \cdot 10^{-2}$ and $0.9 \cdot 10^{-2}$;
            7) $0.2 \cdot 10^{-2}$ and $0.3 \cdot 10^{-2}$.

content, showed that not more than 20-30% of rubber molecules can
be on this surface.  This explains why most of the grafted branches
of polystyrene are inside the dispersed particles.  In connection
with this, the effect of the increase of the degree of polystyrene
grafted onto rubber is to bring to relative growth of the rubber
phase volume at the same polymerization height which was observed
at comparatively low values of BP and $t_1$ (Fig. 1, c, d).  The further increase of BP content and temperature-parameters, which
defines the concentration of free radicals in the system, conditions both the growth of the number of grafted branches and results in considerable decrease in their length.  As a result, as
has already been mentioned in the previous chapter, the total content of the grafted PS becomes smaller at the high temperature
and initiator concentration.  At the same time, this causes some
decrease in the rubber phase volume.  In addition, changes in the
structure of the graft copolymer result  in a decrease in the size
of domains which are formed from the grafted branches of polystyrene.  This may be due to the possibility of a decrease in
compatibility of homopolystyrene macromolecules formed inside the
rubber particles.  Alternatively, a great number of branches makes
the dispersing of rubber solution after phase inversion at equal

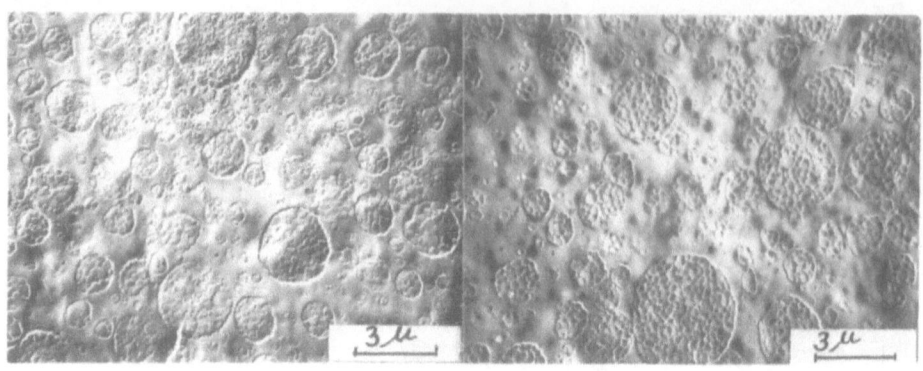

(a)                                        (b)

Figure 3.   Microstructure of HIPS, obtained at varying styrene
            conversion at the first stage of polymerization:
            a)  35%;     b)  55%

intensity of agitation easier, i.e., it promotes decrease of dispersed particles size, which, in turn, results in a decrease of
the fractional volume of the rubber phase in the final product.
This influence of the particle  size on the volume of the rubber
phase is illustrated by the data given in Table II.

        The data on homopolystyrene MW and the content of the graft
copolymer given in Table II show that the agitating conditions at
the stage of polymerization do not significantly influence these
polymer properties.  At the same time, direct interdependence between the size of dispersed particles and the volume of the rubber

| No. | The number of revolutions of the agitator/ min. | Structure | | Graft Copolymer $P_g$, % | $\overline{M_v} \cdot 10^{-5}$ |
|-----|------------------|-----------|-----------|----------------|----------------|
|     |                  | $\overline{V_f}$, % | $\overline{C}$ |          |                |
| 1.  | 80               | 41.3      | 2.16      | 25.8           | 2.84           |
| 2.  | 120              | 36.0      | 1.77      | 24.8           | 2.75           |
| 3.  | 180              | 31.2      | 1.53      | 25.6           | 2.72           |
| 4.  | 250              | 25.8      | 0.95      | 25.5           | 2.35           |
| 5.  | 350              | 24.3      | 0.68      | 24.5           | 2.91           |
| 6.  | 450              | 20.5      | 0.51      | 23.5           | 2.85           |

Table II.   Physical and chemical properties of HIPS obtained by
            the initiated styrene copolymerization with poly-
            butadiene at different intensity of dispersing.

phase is observed - the increase of the particle  size results in
the growth of dispersed phase volume.  This may be explained by
the fact that part of the homopolystyrene molecules, which are
formed in the drops of the rubber phase during the course of polymerization, doesn't succeed in leaving its boundaries because of
the impossibility of overcoming rather large distances to the
interface surface.  Thus, they remain in the occlusions, thereby
raising the rubber phase volume when the size of particles are
large enough (under these copolymerization conditions it is equal
to approximately $0.7\mu$ or more).  The proof for this is the essential difference between the rubber phase volume and the quantity
of the graft copolymer at different sizes of dispersed particles
(Table II).  Hence, at a particle size lower than $1\mu$ , occlusions
are formed by the grafted branches of polystyrene exclusively.
With a highly developed interface surface, an essential part of
copolymer macromolecules is on the surface, with the graft poly-
ştyrene in the matrix, which results in the observed decrease  in
the volume of dispersed particles as compared to the content of
the graft copolymer (Table II, Sample 6).  The further decrease
of dispersed particles size intensifies the above-mentioned effect
even more.  For example, in the case of the particle size equalling
$0.2\mu$. (the polymer formed by the initiated copolymerization in the
presence of the chain length regulator) out of 24% of the formed
graft copolymer, only 11% are included in the volume of rubber
particles.

Analysis of these results shows that optimal correlation of
the values BP and $t_1$, defining the quantity and the length of the
grafted chains when the volume of the rubber phase and the size of
dispersed particles achieve the maximum value, is likely to exist.
In fact, the calculated values of the middle length of the kinetic
chain in the extreme points of dependences of $\overline{V}f$ on BP and $t_1$, are
approximately equal (Table III) for the conditions in question.

| Fixed levels of variables | For dependence of $\overline{V}f$ on BP | | For dependence of $\overline{V}f$ on $t_1$ | |
|---|---|---|---|---|
| | $BP \cdot 10^2$ mol/1: | | $t_1$, $^\circ C$ | |
| - 1 | 1.4 | 440 | 85 | 670 |
| 0 | 0.8 | 420 | 80 | 500 |
| + 1 | 0.4 | 440 | 75 | 410 |

Table III.  The value of mean length of kinetic chain at the
extreme points of dependence of $\overline{V}f$ on BP and $t_1$.

The extreme character of the dependence of the dispersed particle size on BP and $t_1$ (Fig. 1 - 1,2,3) can be explained by the influence on the dispersion not only of the quantity and structure of the graft copolymer, but of the relation of the viscosities of the emulsion phases (9). At low values of the investigated parameters, when the viscosity of the polystyrene phase is essentially the main cause for high MW, this factor proves to be the defining one more than the fact that smaller drops are being formed. At high concentrations of BP and high $t_1$, as has already been mentioned, the main part is played by the structure of the intraphase layer as influenced by the copolymer structure.

The latter also affects the structure of polystyrene occlusions, i.e., instead of the usually formed cellular rubber particles, the "tangle" particles and the particles of "capsule" type are formed (Figure 4). Analogous changes of particle types were observed by Ekhte when investigating the structure of a homopolystyrene mixture with block copolymers of different block size (9).

The data presented here allow us to conclude that the initiator concentration and polymerization temperature affects HIPS morphology by influencing the quantity and structure of the graft copolymer.

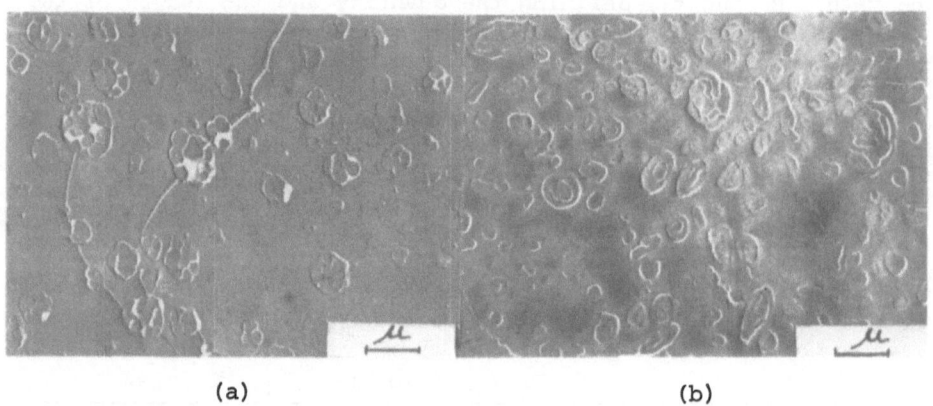

(a)                                        (b)

Figure 4.  Morphology of HIPS samples obtained by styrene copolymerization with rubber under the following conditions:
a) $t_1 = 80^{\circ}C$;    $t_2 = 115^{\circ}C$;    BP = 0;    TBBP = $0.6 \cdot 10^{-2}$mol/l
b) $t_1 = 90^{\circ}C$;    $t_2 = 125^{\circ}C$;    BP = $1.0 \cdot 10^{-2}$mol/l; TBBP = $0.9 \cdot 10^{-2}$mol/l.

## REFERENCES

1. M. Baer, J. Appl. Polym. Sci., 16, 1109 (1972).
2. G. Cigna, J. Appl. Polym. Sci., 14, 1781 (1970).
3. G. Cigna, S. Matarrese, and G. Biglione, J. Appl. Polym. Sci., 20, 2285 (1976).
4. G. Riess, S. Marte, J. L. Refregier and M. Schlienger, Polym. Sci and Techn., 10, 327 (1977).
5. V. D. Yenalyev, N. A. Noskova and S. Hohne, ACS Polym. Prepr., 37, 575 (1977).
6. V. D. Yenalyev, N. A. Noskova and B. W. Kravchenko, "Emulsion Polymerization," ACS Symposium Series, Vol. 24, p. 379, ACS, Washington (1976).
7. V. D. Yenalyev, N. A. Noskova and V. I. Melnichenko, Polym. Prepr., 37, 645 (1977).
8. G. D. Ballova, V. M. Bulatova, K. A. Vylegzhanina, E. I. Yegorova, L. L. Sulzhenko and G. P. Fratkina, Vysokomol. Soed., A 11, 1827 (1969).
9. A. Echte, Angew. Macromol. Chem., 58/59 (1977).

# STUDIES OF STYRENE COPOLYMERIZATION WITH RUBBER. III. THE PROCESS

# PECULIARITIES BROUGHT ABOUT BY THE PRESENCE OF ALIPHATIC MERCAPTANS

V. D. Yenalyev, V. I. Melnichenko, N. A. Noskova,
O. P. Bovkunenko, A. N. Shelest

Donetsk State University
Donetsk, 340055, USSR

The regulating action of aliphatic mercaptans in free radical polymerization has been known for a comparatively long time. The reaction rate constants for chain transfer for some mercaptans were found as well as the conditions for their effective use when obtaining different polymer composites, including high impact polystyrene (HIPS). However, introduction of transfer-agents (regulators) into polymerization systems affects not just the molecular weight characteristics of the polymer. In addition to lowering the MW of homopolystyrene which is formed during styrene polymerization with polybutadiene, mercaptans change the conditions of rubber phase dispersion and thereby affect the HIPS morphology. The extent of this influence is reflected by the regression equations, which we developed for the conditions of thermal styrene copolymerization with rubber in the presence of n-lauryl mercaptan according to a three-staged temperature-time regime (4):

$$\overline{Vf} = 25.6 + 9.3X_1 + 1.3X_3 + 2.4X_4 + 0.6X_1X_3 - 3.1X_1X_4 -$$

$$0.7X_2X_4 \tag{1}$$

$$\overline{C} = 3.4 + X_1 - 0.3X_2 + 0.1X_3 + 1.4X_4 + 0.3X_1X_4 - 0.4X_2X_4 +$$

$$0.1X_2X_3 \tag{2}$$

where:  $\overline{Vf}$ is the relative volume of the rubber phase, %; $\overline{C}$ is
the mean size of dispersed rubber particles; $X_1$ is the
polybutadiene content ($6 \pm 2\%$);  $X_2$ is the polymerization
temperature at the first stage ($120 \pm 10^{\circ}C$);  $X_3$ is the
polymerization extent after the first stage completion
($50 \pm 15\%$);  $X_4$ is the concentration of n-lauryl mercaptan

$(6.73 \pm 4.49 \times 10^{-4}$ mol/1).

Polymerization temperatures at the second and third stages were not varied (150°C and 190°C respectively). Monomer conversion after the completion of the second stage was maintained at 88-90% for all regimes. Without giving a detailed analysis of equations (1) and (2), we would like to point out that regulator concentration affects the size of particles in the greatest way among the considered factors. Hence, the influence of both regulators and initiators on the polymerization process is of a complicated character and is different at different stages. Consequently, it is extremely difficult to predict the properties of the HIPS formed in the process, especially in the presence of both regulators and initiators. In fact, the literature (2,3) points out the considerable diminishing of the rate of initiated styrene polymerization in the presence of n-lauryl mercaptan, n-dodecyl mercaptan and t-dodecyl mercaptan as regulators. The probable course of the chain termination step has been postulated as follows (2,3):

$$R_m + C_nH_{2n+1}S \longrightarrow C_nH_{2n+1}SR_m \qquad (1)$$

$$2C_nH_{2n+1}S^{\cdot} \longrightarrow C_nH_{2n+1}SSC_nH_{2n+1}SSC_nH_{2n+1} \qquad (2)$$

We do not feel this explanation of the peculiarities of polymerization to be sufficient since re-combination reactions in the presence of primary radicals, formed from the regulator molecules, takes place in thermal polymerization of styrene as well. However, in this case the change of polymerization rate with increase of n-lauryl mercaptan concentration wasn't noted (4). In connection with this, we feel it expedient to consider styrene polymerization as a simpler system in the presence of widely used peroxide initiators (e.g., benzoyl peroxide (BP), t-butyl perbenzoate (TBPB), and regulators - n-lauryl mercaptan (NLM), n-butyl mercaptan (TMB), t-octyl mercaptan (TOM), t-decyl mercaptan (TDM), t-dodecyl mercaptan (TDDM), and t-pentadecyl mercaptan (TPDM)).

The dependence of styrene conversion on polymerization time in the presence of BP and NLM shows that increase in the amount of regulator with constant primary initiator concentration results essentially in decrease of the polymerization rate. Decrease of polystyrene MW is observed only at the initial stage of process. With the increase of monomer conversion the polystyrene MW is increasing. The higher the initial NLM concentration, the faster it achieves the MW obtained under the same conditions but without the regulator at a styrene conversion equal to 10-30%. These trends are characteristic for styrene polymerization in the presence of BP and TBPB with the other aliphatic mercaptans as well. Figure 1 shows that decrease in the length of the t-mercaptan results in lowering of the reaction rate.

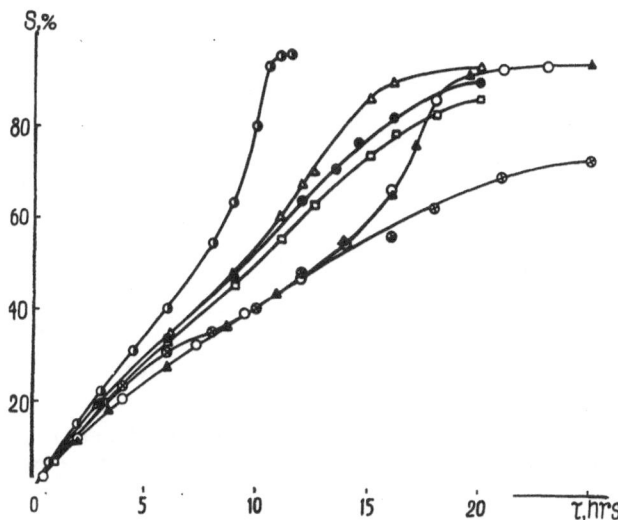

Figure 1.  Kinetics of initiated styrene polymerization in the
presence of mercaptans.  Temperature is 75°C; BP concentration if 1.26 x 10$^{-2}$ mol/l; the concentration of
mercaptans is 0.63 x 10$^{-2}$ mol/l.

◖ ⊤ BP;          Δ - BP + TPDM;     ● - BP + TDDM;
□ - BP + TDM;    ▲ - BP + NLM;      ○ - BP + NPM'
⊗ - BP + TOM.


        Obviously, introduction of the regulator into the polymeri-
zation system results in change of kinetic regularities of initiated styrene polymerization and doesn't decrease the MW of polystyrene which is formed at the high stages.  It is difficult to
explain the obtained regularities by the leakage in the course of
polymerization of reactions of the types 1 and 2, since in this
case the increase of polymer MW with increase of styrene conversion cannot be explained.  The presence of mercaptans can be postulated to result in either lowering of the peroxide concentration
because of their direct interaction or lowering of its initiating
activity.  To clear up the character of this interaction, model
mixtures of NLM, NBM, TDDM in benzene and styrene were investigated
both at room temperature and at higher temperatures (up to 60°C).
Figure 2 gives dependences of BP concentration on the ratio NLM:BP
and regulator concentration on the ratio initiator:regulator, the
curves character depending neither on duration of reagents contact
(from 10 minutes to 10 hours) nor on mixture temperature (from
20 to 60°C).  The presented relationships show that the peroxide
interacts with mercaptan before polymerization begins, the reaction
between them not taking place to completion, because even a great
excess of one of the components doesn't result in full disappear-
ance of the other.

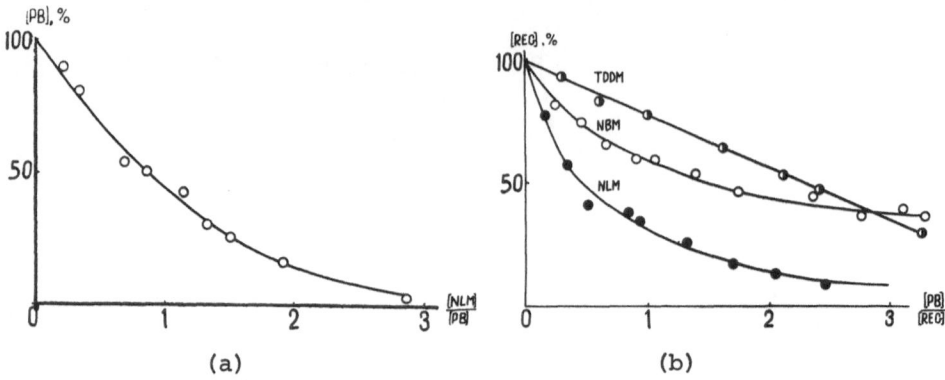

Figure 2.   Dependence of remaining BP concentration on the molar
            ratio NLM/BP (a) and remaining regulator concentration
            on the molar ratio BP/regulator.

          Mercaptans are known (5) to become oxidized by oxygen and
peroxy compounds in the absence of a catalyst.  The main products
of this reaction are disulfides of the corresponding mercaptans
and organic acids formed from initiator fragments.  Thus, the
joint presence of peroxy initiators and aliphatic mercaptans in
the polymerization system results in the appearance of new components which may affect both the polymerization rate and the MW of
the formed polymer.  To elucidate the influence on thermal styrene
polymerization in the presence of n-lauryl disulfide (NLD) and benzoic acid (BA), the products of the interaction of BP with NLM were
investigated.  Neither BA nor NLD prove to affect the rate of thermal styrene polymerization.  No essential influence of these substances was found on the MW of polystyrene formed either at the
initial or high stages of the process.  Thus, we concluded that
neither BA nor NLD have properties of initiators, regulators, or
inhibitors of free radical polymerization.  The peculiarities of
styrene polymerization are likely to be connected with the simultaneous presence of NLM with BP, NLD, BA and the other components.
To get more detailed information about the mechanism of the styrene
polymerization process up to high stages, polymerizations were
carried out in the presence of the following mixtures:  1) BP
with BA;  2) BP with BA and NLD;  3) BP with NLD, the initial ratio
of the initiator and regulator concentrations being 2:1 in the
latter case.  At this ratio, approximately 90% of mercaptan will
interact with the initiator before the beginning of polymerization
(in accordance with Figure 2), forming the model system close to
the second (BP with BA and NLD).  The obtained dependences of styrene conversion on time and polymer MW on styrene conversion for
the corresponding conditions are presented in Figure 3.

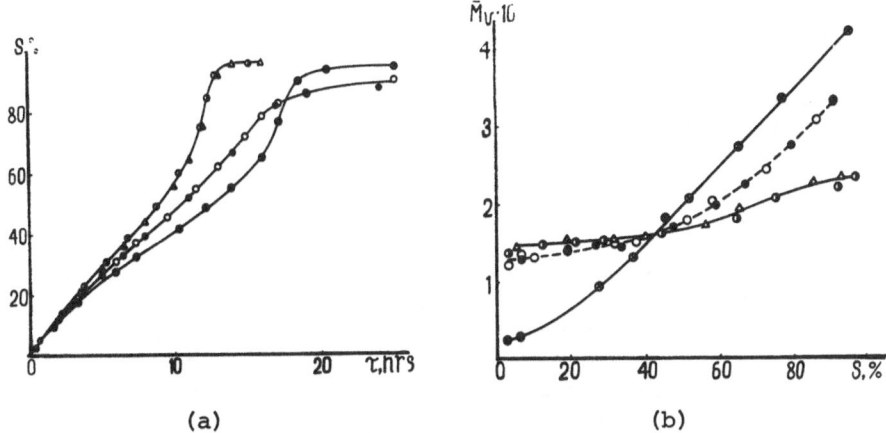

(a)                                          (b)

Figure 3.   Dependence of the styrene conversion on time (a) and
            polystyrene MW on conversion (b), obtained for the
            following conditions:  polymerization temperature is
            75°C;  components concentration is:  $\mathbf{0}$ - $0.9 \cdot 10^{-2}$
            mol/l BP;  $\Delta$ - $0.9 \cdot 10^{-2}$ mol/l BP, $0.6 \cdot 10^{-2}$ mol/l BA;
            $0$ - $0.9 \cdot 10^{-2}$ mol/l BP, $0.6 \cdot 10^{-2}$ mol/l BA, $0.3 \cdot 10^{-2}$
            mol/l NLD;  $\bullet$ - $0.9 \cdot 10^{-2}$ mol/l BP, $0.3 \cdot 10^{-2}$ mol/l NLD;
            $\circledast$ - $1.2 \cdot 10^{-2}$ mol/l BP, $0.6 \cdot 10^{-2}$ mol/l NLM.

        Analysis of curves 1 and 2 in Figure 3-a,b results in the
conclusion that in the case of styrene polymerization initiated by
BP, BA doesn't affect either the process rate or polymer MW.  These
data also prove that introduction of NLD into the polymerization
system considerably lowers the process rate at the initial and
high stages, although it has no inhibiting action, as was stated
when investigating thermalstyrene polymerization.  The data are in
agreement with the results of the studies of the interaction of BP
with methyl sulfides, t-butyl sulfide and the disulfide of t-butyl
and t-amyl (6).  Here it was stated that at temperatures of 70-80°C,
the reaction between BP and sulfides proceeds according to a nonrad-
ical mechanism with formation of benzoic anhydride, benzoic acid,
olefins, different sulfoxides and other products.  Thus the lowering of the process rate as compared to the polymerization rate in
the absence of disulfide can be explained by reaction between BP
and NLD, which results in lowering of initiator concentration in
the system.  However, this lowering of polymerization rate doesn't
result in the expected increase of polymer MW at the initial stage
of reaction (see curves 1 and 2 in Figure 3-b).  The products being
formed in the interaction of peroxide with disulfide are likely to
have the properties of weak inhibitors.  With the increase of polymerization to high conversion, owing to diffusion control of the
process of macroradicals interaction, the role of reactions with
participations of small molecules increases, which leads to

intensification of weak inhibitor action. Investigations showed
that the increase of initial NLD concentration causes gradual lowering of polymerization rate at the high stages and gel effect
characteristics of initiated styrene-polymerization are not observed. Significant increasing of MW of the polystyrene being
formed at the high stages is not observed either as compared to
the polymerization process in the absence of NLD, although the
rates of both processes considerably differ (see curves 1 and 2 in
Figure 3-a,b).

Figure 3 also shows the dependences of the styrene conversion
on time and polystyrene MW on styrene conversion in the presence
of BP and NLM. Their analysis allows us to conclude that the
reaction between BP and NLM doesn't take place to completion in
reality. In the polymerization system there is a regulator, although the initiator was introduced with a considerable excess as
compared to the supposed stoichimetric ratio of their interaction.

The data allows us to propose the following mechanism of
styrene polymerization in the presence of peroxide initiators and
aliphatic mercaptans: the regulator interacts with the initiator
forming organic acids, disulfides and other products; disulfides
in their turn begin interactions with the initiators but at a
lower rate than mercaptans; this reaction is proceeded by the faster consumption of the benzoyl peroxide in the presence of different
mercaptans, but there was no increase of radical formation rate.
Due to the proceeding of these processes in the system, substances
are being accumulated which have the properties of weak inhibitors.
Their quantity is limited by the consumption of the regulator by
a chain transfer, which is why their inhibiting action at the high
stages is not observed in the same way as was seen in styrene polymerization in the presence of BP with disulfide (the latter doesn't
take part in the chain transfer reaction for all practical purposes
and its consumption is connected only with the interaction with the
initiator). Fast consumption of initiator and regulator results
in significant increase of the polystyrene MW at the high stages
of polymerization. The relative increase in polymer MW depends
on the initial ratio initiator/regulator. Dependences of polystyrene MW and homopolystyrene HIPS matrix on NLM concentration
at constant BP concentration (0.18%) are presented in Figure 4.

Variation of the ratio initiator/regulator occurred after
25% formation of emulsion structure in HIPS synthesis, thus influence of the interaction of the initiator with the regulator
was excluded. Curves 1 and 2 differ in temperature of styrene
polymerization at the first stage. The presence of maxima in
the curves may be explained by the following (when the ratio of
peroxide/regulator is high (more than 4)). For all practical purposes there is no NLM in the polymerization system (see Figure
2-b) and the increase in polystyrene MW is caused by the consump-

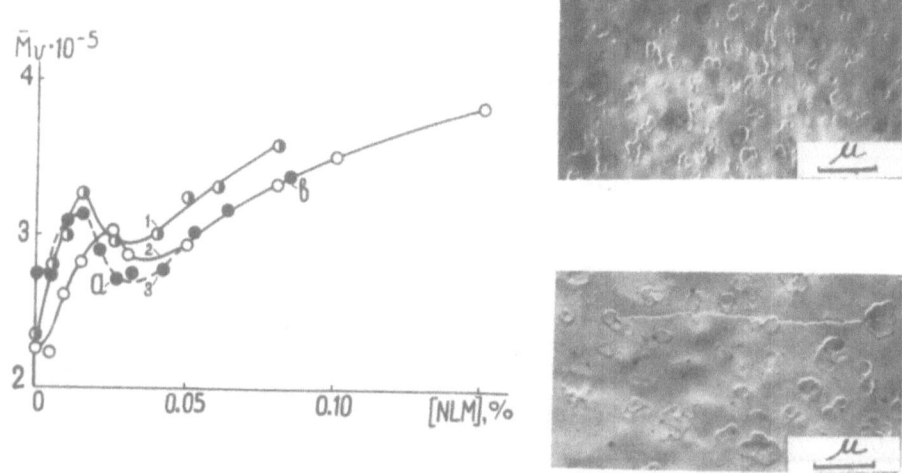

(b)

Figure 4.   Dependence Mv of polystyrene and HIPS morphology on
            ratio NLM/BP;  the structure is presented for the
            samples (a) and (b).


tion of the initiator by interaction with the regulator.  The increase of NLM concentration, i.e., the decrease of the ratio init-
iator/regulator, causes noticeable appearance of mercaptan in the
system and it brings about lowering of homopolystyrene MW.  Further increase of regulator concentration causes significant lowering of peroxide quantity in the system (Figure 2-a).  As a result,
the regulating action of NLM to MW will be hidden by the formation
of a high molecular weight fraction of polymer at the high stages.
The increase in polystyrene MW at a low ratio initiator/regulator
is explained by this.  Polybutadiene presence in the system doesn't
change the character of this dependence and causes only the shift
of the maximum (Figure 4, curve 3).

     The ratio initiator/regulator affects not only the polymer-
ization rate and polymer MW, but also the process of copolymer formation, and graft branch MW, and thus results in changes to the
HIPS microstructure formed at the initial stages.  The graft polystyrene branches with high MW promote formation of large polystyrene domains inside the rubber particles, thus increasing both
the particle  size and the relative volume of the rubber phase.
This data is presented in Table I and Figure 4.  Table I shows that
good correlation between the relative volume of rubber phase and
homopolystyrene matrix MW is observed.  These results allow us to
understand the mechanism of the polymerization process in the
presence of peroxide initiators and mercaptans in greater depth.

Their usefulness also lies in the fact that one can choose a good formulation of polymerization components at the given regime of reaction and predict the proceedings of the whole process.

| NLM, % weight | 0 | 0.015 | 0.026 | 0.052 | 0.084 |
|---|---|---|---|---|---|
| $Mv \cdot 10^{-5}$ | 2.75 | 3.14 | 2.70 | 3.18 | 3.40 |
| $\overline{Vf}$, % | 10.8 | 12.5 | 10.1 | 12.4 | 14.4 |
| $\overline{C}$ | 0.22 | 0.31 | 0.19 | 0.26 | 0.32 |

Table I.   The values $\overline{Vf}$, $\overline{C}$ and Mv of HIPS obtained at different NLM concentrations.

REFERENCES

1. R. A. Gregg, D. M. Alderman, F. R. Mayo, J. Amer. Chem. Soc., 70, 3740 (1948).
2. V. D. Yenalyev, T. N. Sadovskaya, V. V. Zaitseva, "Synthesis and Physical Chemistry of Polymers," p. 29, Naukova dumka, Kiev (1966).
3. V. F. Kazanskaya, S. V. Smirnova, Vysokomol. Soed., B XII, 523 (1970).
4. V. D. Yenalyev, V. I. Melnichenko, O. P. Bovkunenko, "Thermal Polymerization of Styrene in the Presence of n-Lauryl Mercaptan, Dep. VINITI, No. 612-77 Dep.
5. Sigeru Oae, "Chemistry of Sulfur Organic Compounds, Chemistry," Moscow (1975).
6. A. W. Pryor, H. T. Bickley, J. Org. Chem., 37, 2885 (1972).

TECHNIQUES OF BLOCK COPOLYMER PREPARATION

G. Allan Stahl

BF Goodrich Research and Development Center
9921 Brecksville Road
Brecksville, OH  44141

ABSTRACT

The unusual physical properties of block copolymers range
from combination of rubbery and glassy polymers to polar
and non-polar polymers. The uses of these block copolymers thus
range from thermoplastic elastomers to surfactants to modifiers.
It is therefore necessary that the polymer chemist stay abreast of
their many preparative techniques. The purpose of this report is
to classify these techniques, serve as a reservoir of literature
references, and as a source of ideas to stimulate thinking on
particular block copolymer preparations and their potential
utility.

The block copolymer preparative techniques reviewed in this
report include sequential addition to macroanions, "pseudoliving"
carbenium ion systems, used of coordination catalysts, coupling of
polymer termini, and various preparations employing free radical
polymerization.  The review covers 133 papers and patents.

INTRODUCTION

The unusual physical properties of block copolymers have
enticed many workers to perform a tremendous amount of research
since Szwarc first demonstrated sequential anionic addition in
1956[1].  A vast and growing literature has evolved, and in it are
a number of excellent reviews of block copolymer synthesis and

properties (2-8). In these reviews block and graft copolymers have been treated as a single topic; however, with the increased sophisti-cation of preparation and application, block copolymers have become a subject worthy of individual study.

This report reviews the major synthetic approaches by canvassin; the literature of block copolymer preparation. Discussion of the physical properties of these materials, or their application in physical blends and alloys is beyond the scope of this report. The major purpose of this work, thus, is to provide a platform for the preparation of block copolymers, and to stimulate ideas for new preparative techniques and applications for the existing techniques.

Literature references range from very early work in block copolymer synthesis to some recent techniques, yet due to the volume of papers, a comprehensive review is not intended. The actual search was conducted by computer using key-work references. Each reference listed by the computer (and there was more than one-thousand) was reviewed, and accepted or rejected by qualitative judgement of tis appropriateness in this review.

## Preparation of Block Copolymers by Anionic Polymerization

Anionic polymerization is the preferred technique to produce block copolymers free of the respective homopolymers. Discovered by Szwarc[1], the "living" anionic polymerization differ from other polymerization systems in that it lacks a termination step. The absence of termination permits the preparation of nearly pure block copolymers of known molecular weight and composition. The prepara-tion of block copolymers anionically does have two major drawbacks. Ultrapure reagents are required, and only a few vinyl monomers polymerize well by this technique.

The most extensively studied block copolymers prepared by anionic polymerization are the styrene-butadiene or styrene-isoprene rubbers. Shell Chemical Company's Kraton® thermoplastic elastomers are ABA block copolymers of this type. Their elastomeric properties are excellent, yet they differ from other rubbers in that vulcani-zation is not required. These elastomers consist of a rubbery polybutadiene matrix with the styrene segments serving as anchors in thermoplastic microdomains.

Preparation of styrene-diene block copolymers involves sequen-tial addition of the diene monomer to "living" polystyrenes. The resulting polymer is an AB block copolymer consisting of segments of styrene mers and diene mers.

A large number of patents and papers have been published since 1956 covering this technique(9-13). The initiator in most cases is an organolithium compound.  Organosodium or organopotassium compounds can be used, but a much higher amount of the 1,2 addition product of the diene is found in the product.

Difunctional initiators such as sodium naphthalene can be employed to prepare triblock ABA block copolymers.  Difunctional initiators produce "living" polymeric dianions which are capable of adding a second monomer at each end.  On the addition of a second monomer the ABA structure results.  For example, consider the preparation of poly(butadiene-b-styrene-b-butadiene) by anionic polymerization initiated by sodium naphthalene.

The radical-anions couple, producing dianions.  It is the dianions which is the initiating species as shown below.

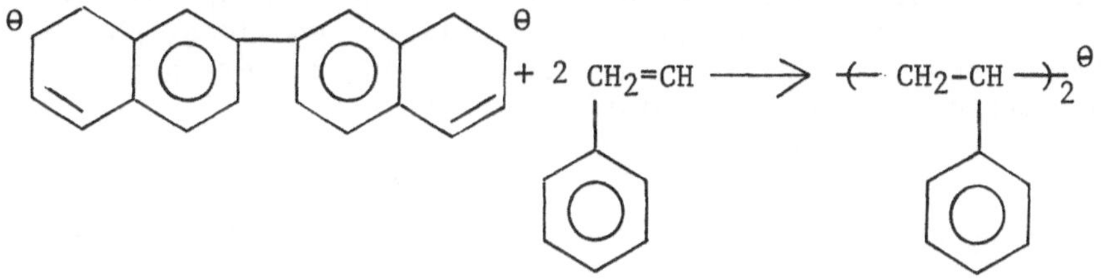

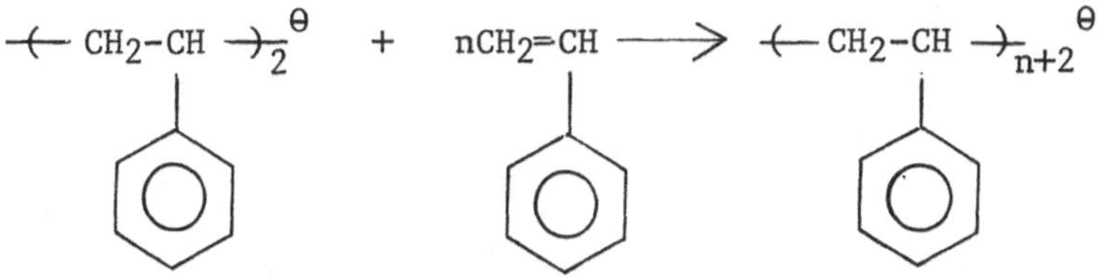

When butadiene is added an ABA block copolymer is formed.

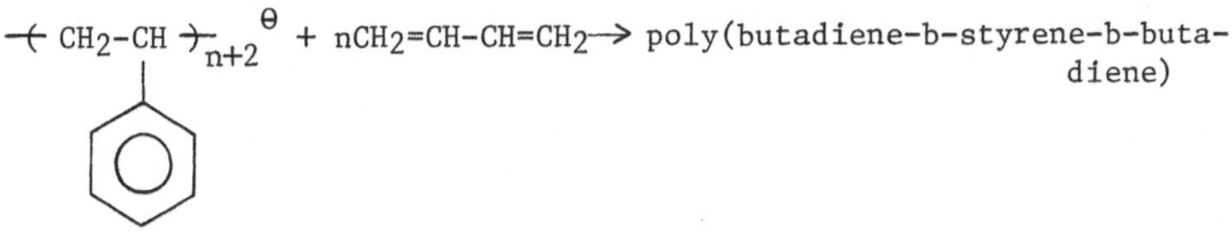

   Polyfunctional polybutadienyldilithium anions have been used
to prepare ABA block copolymers of butadiene-styrene[14,15] and butadiene-α-methyl styrene[16]containing rubbery center blocks.  Comparable block copolymers of butadiene-α-methyl styrene with α-methyl
styrene center blocks have also been reported[19,20].In a similar
preparation, Worsfold has capped an isoprene-α-methyl styrene block
copolymer with p-divinyl benzene[21].This polymer is capable of
further copolymerization.

   Workers have reported the use of silicon tetrachloride to
prepare 3 or 4 arm star-block copolymers of butadiene-styrene[17,18]
Silicon tetrachloride terminates several living macroanions at a
single junction by halogen exchange.  A 4-arm star block copolymer
is represented below.

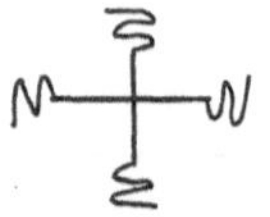

A 4-arm star-block copolymer

Multiarm star-block copolymer can be prepared by incorporation of
a small amount of divinylbenzene.

   AB and ABA block copolymers may also be prepared by coupling
the "living" anion with a suitable anion reactive coupling agent.
The coupling agents include methylene halides, epoxides, phosgene,
as well as some anhydrides.  The Shell Kraton® rubbers are prepared
in this manner.  Methylene chloride, as shown below, is employed to
couple the AB block macroanions.

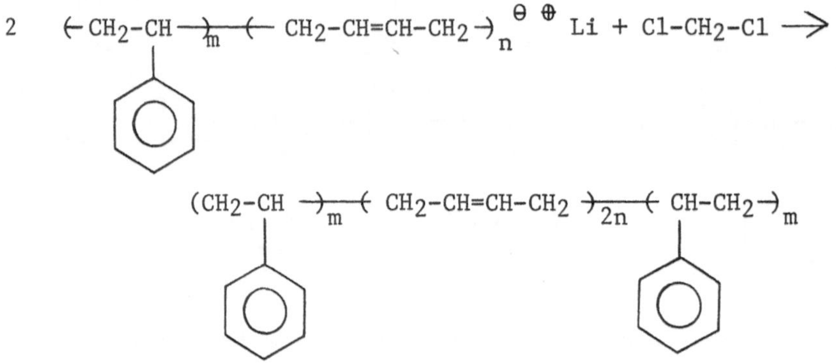

The resulting triblock polymer is an ABA structure of controlled
segment molecular weights.  Anion coupling has been employed in the
preparation of block copolymers of styrene-chloromethyl styrene[23]
and styrene-p-fluoromethyl styrene[25].Styrene-perfluoroethylene
oligomers have also been prepared by anion coupling[26].  Other

workers have used a slightly different type of anion coupling when
they warmed "living" styrene macroanions in the presence of poly-
(tetrahydrofuran) cations to produce block copolymers(22,24).

It is obvious that anion chain transfer must be avoided to
prepare good yields of block copolymers with "living" macroanions.
Because of susceptibility to chain transfer many monomers capable of
anionic polymerization will not readily produce block copolymers.
For example, methyl methacrylate block copolymers are much less
studied than those of styrene.  Anion chain transfer occurs at the
pendent ester group, drastically reducing the yield of block copolymers.  Poly(methyl methacrylate-b-isoprene) has been prepared,
however, by using an ingenious chain cap of 1,1'-diphenylethyl-
ene(27,28). 1,1'-diphenylethylene will not anionically homopolymerize,
therefore it adds only one mer to the macroanion.  This anion is
more stable in the presence of methyl methacrylate, but will
initiate further polymerization.  Other workers have reported the
preparation of isoprene-methyl methacrylate block copolymers by
sequential addition to "living" polyisoprene anions(29,30).

Anionic polymerization of lactams offers the best approach to
the preparation of polyamide containing block copolymers.  Styrene-
nylon 6 block copolymers were prepared by adding ε-caprolactam to
polystyrene macroanions terminated with bisphenol A bis(chlorofor-
mate)(31). Yamashita prepared ABA block copolymers of styrene-α-
pyrrolidone and styrene-ε-caprolactam by sequential addition to
styrene macroanions(32). Similarly Stehlik and Sebenda prepared
N-acrylamide containing block copolymers(33). Block copolymers of
isoprene-pivalolactam have also been reported(34-36). In these cases
the lactam was added to "living" polyisoprene anions.

Block copolymers consisting of segments with widely separated
solubility characteristics have generated considerable interest because of their unusual surfactant properties.  In fact, one of the
earliest commercial block copolymers were the Wyandotte "Pluronics."
These were poly(propylene oxide-b-ethylene oxide) prepared by sequential addition of ethylene oxide to sodium alkoxide initiated
propylene oxide (37,38).  Szwarc (39) and others (40,41) prepared
poly(styrene-b-ethylene oxide) by addition of ethylene oxide to
polystyrene anions in tetrahydrofuran.  Other syntheses of AB or ABA
block copolymers of styrene-ethylene oxide include sequential addition in various solvents, and coupling reactions (42,43).

Similar block copolymers have also been prepared by coupling
methyl methacrylate macroanions and suitably terminated poly(ethylene
oxide)(44)and by the initiation of methyl and ethyl methacrylate by
alkali metal alkoxides of poly(ethylene oxide)(45). Styrene containing
block copolymers have also been produced(46). In the former plus
latter techniques the ester containing block segments were subsequently hydrolyzed to methacrylic acid thus making them water soluble.

Other interesting block copolymers have been prepared recently by sequential monomer addition to "living" macroanions. Hale and Pope report the preparation of an ABA block copolymer of ethylene sulfide-isoprene using lithium naphthenide as the initiator[47]. Gallot has described the synthesis of poly(butadiene-b-vinyl naphthalene)[48] while Szwarc has prepared p-xylylene block copolymers containing vinyl pyridine or styrene. These latter copolymers are prepared by producing p-xylylene vapor in the presence of "living" polystyrene or vinyl pyridine anions, or by the novel reaction of vinyl pyridine with p-xylylene radicals[49]. Block copolymers of ferrocenylmethyl methacrylate have also been prepared[50]. This polymer is particularly interesting in light of the growing application of organometallic polymers as organic reaction catalyst.

## Preparation of Block Copolymers by Cationic Polymerization

Since there is no known "living" carbenium ion system, block copolymer synthesis by sequential addition is unknown. Block copolymers have, however, been produced by cationic polymerization. The preparation is based on Kennedy's discovery that certain alkyl-aluminum compounds ($Et_2AlCl$ for example) will initiate the cationic polymerization of certain active vinyl monomers in the presence of alkyl halides. By this technique, a halide terminated polymer can, in the presence of $Et_2AlCl$, initiate polymerization of a second monomer to form a block copolymer.

Kennedy has prepared poly(styrene-b-isobutylene)[51,52,53] by this technique employing 2-bromo-6-chloro-2,6-dimethyl heptane as the initiator. Similarly, Stannett has prepared poly(vinyl carbazole-b-isobutyl vinyl ether) by addition of the vinyl ether to poly(vinyl carbazole) initiated with $Ph_3CSbCl_6$[54]. To successfully prepare block copolymers it is necessary to maintain a low solvent/monomer ratio (to surpress chain transfer), and a high initiator concentration. The production of block copolymers by this technique has been called "psuedo living" carbenium ion polymerization.

## Preparation of Block Copolymers Using Coordination Catalysts

Stereospecific catalysts, such as the Ziegler-Natta type catalysts, have been employed in the preparation of block copolymers. It is obvious that the growing chains must have a long active lifetime or block copolymerization will not be possible. At present, the lifetime of a growing polyethylene chain has not been determined to everyone's satisfaction, but most agree that a catalyst based on $TiCl_2$ and $Et_2AlCl$ produce growing chains with a life of several minutes.

Most of the literature covering ethylene-propylene block copolymers are process patents[55,56], however, specific papers on the subject do exist. For example, it has been reported that a block copolymer of 4-methyl pent-1-ene and propylene was prepared using a

$TiCl_2 \cdot Et_2AlCl$ catalyst[57,58]. Using this catalyst, or a similar catalyst, block copolymers of the following monomer pairs have been prepared:  1-butene-ethylene[59], 1-butene-propylene[60], ethylene-styrene[61] α-methyl styrene-ethylene[62] and propylene-vinyl cyclo-hexane[63].

More polar monomers have also been incorporated in block copoly-mers with ethylenic monomers.  In one case, polypropylene was capped by dimethylaminoethyl methacrylate[64] while in another polyethylene was capped by methyl methacrylate[65]. In these polymerizations $TiCl_3 \cdot Et_2AlCl$ and benzoyl peroxide were the initiators.  The peroxide presumably oxidizes the metal alkene bond to a metal salt of a macrohydroperoxide.  The macrohydroperoxide then is used to initiate radical polymerization of the methacrylate monomer.  Similarly a polypropylene-b-acrylonitrile material has been produced[66]. A $TiCl_4 \cdot AlCl_3 \cdot Et_2AlCl$ catalyst was employed to produce polypropylene.  Treat-ment of this polymer with ammonia and oxygen in turn produced a macrohydroperoxide, which subsequently initiated block copolymeriza-tion of the acrylonitrile.  A block copolymer of potentially inter-esting properties, poly(butadiene-b-propylene oxide) has been prepared using a $Bu_3Al \cdot TiCl_3$ catalyst.

New block copolymers of polystyrene and poly(ε-caprolactone) have recently been prepared with a combination of coordination catalyst and anionic polymerization[67,68]. An amorphous styrene block was prepared by conventional anionic polymerization at room tempera-ture, and terminated with ethylene oxide.  After hydrolysis the terminal – OH groups were substituted by oxo–Al–Zn alkoxide.  A crystalline block was then added by coordination polymerization of ε-caprolactam.

## Preparation of Block Copolymers by Coupling of Polymer Termini

Block copolymers may be formed if suitable reactive end groups of two polymers can react together or react with a linking molecule. All basic types of block copolymers are possible by this technique, including AB, ABA, and sequential ABAB... block copolymers.  Suitable linking molecules include diisocyanates, phosgene, and isophthaloyl chloride.  For example, isocyanates readily react with functional groups containing active hydrogens (i.e. –OH, $–NH_2$, –COOH).  Two polymers terminated with functional groups containing active hydrogen may thus be linked covalently by a diisocyanate.  Nylon 6 and poly-(tetramethylene glycol) were linked in this manner by 2,4-toluene diisocyanate to form a block copolymer[69].

In other applications, Baysal utilized methylene dicyclohexyl isocyanate to link polybutadiene with polystyrene or poly(methyl methacrylate) to form the appropriate block copolymers[70]. Diphenyl methane-4,4'diisocyanate was similarly used to prepare a block

copolymer of cellobiose and poly(diethylene glycol adipate)[71]. This
latter block copolymer is particularly interesting when the high
availability and low cost of cellobiose is considered.

Hergenrother and Ambrose have prepared butadiene-imide[72] and
butadiene-caprolactam[73,74] block copolymers using polybutadiene
capped on each terminus by isocyanate groups.  Poly(ethylene-propylene adipate-b-methyl methacrylate) was prepared by heating hydroxy
terminated poly(methyl methacrylate) with the isocyanate terminated
reaction products of hydroxy terminated poly(ethylene-propylene
adipate) and bis(4-isocyanatocyclohexyl)methane[75].

Yamashita combined amidine terminated poly(styrene-co-acrylonitrile) and poly(tetramethylene glycol) with phosgene, to form a
sequential block copolymer[76]. Similarly in a patent assigned to
General Electric, White reported preparing a phenylene oxide-car-
bonate block copolymer by heating hydroxy terminated poly(phenylene
oxide) and a polycarbonate in the presence of phosgene[77]. Other
block copolymers containing segments linked by a molecule of phosgene
include poly(butadiene-b-styrene) and poly(isoprene-b-styrene)[78].
These materials were reported prepared by the reaction of hydroxy
terminated polydiene and polystyrene pretreated with phosgene.

Isophthaloyl chloride has also been employed in block copolymer
preparation.  Poly(m-phenylene isophthalamide) and either poly-
(ethylene oxide) or poly(dimethyl siloxane) block copolymers have
been prepared by terminating the former with isophthaloyl chloride
and subsequently adding hydroxy terminated poly(ethylene oxide) or
poly(dimethyl siloxane)[79]. Block copolymers were also produced when
a polyformal was capped with 1,6-hexanediol, and then heated in the
presence of poly(tetramethylene terephthalate)[80]. In another interesting application, polyesterpolyether block copolymers of poly-
(ethylene terephthalate) and a number of polyglycols have been
reported[81]. Their synthesis involves esterification, catalyzed by
zinc acetate/titanyl oxalate.  ABA block polymer of 2-hydroxyethyl
methacrylate and styrene was prepared by coupling -NH$_2$ terminated
PHEMA and isocyanate terminated polystyrene.  The latter material
was prepared by UV polymerization of styrene in the presence of
bis(p-isocyanatophenyl)disulfide[82].

## Preparation of Block Copolymers by Free Radical Polymerization

Introduction.  We have, so far, considered ionic propagation,
coordination catalysis, and the step reactions of a polymer terminus
as techniques for the preparation of block copolymers.  Free radical
polymerization may also be utilized by application of one of several
chemical manipulations.  For example, block copolymers may be prepared by coupling macroradicals, or by generating new radicals in
the presence of a second monomer by photolytic or mechanical degradation.  As an alternate, difunctional initiators may be employed.

These are usually diperoxide or peroxide-azo containing molecules, and thus can be decomposed in two steps by temperature control. The first cleavage initiates polymerization and production of a macro-peroxide or macroazo molecule. The second cleavage is in the presence of a second vinyl monomer, and a block copolymer results. A last, yet much less clearly defined technique of block copolymer preparation, will also be reviewed. Specifically, block copolymers may be prepared by addition of a second vinyl monomer to occluded "living" macroradicals. Thus by use of the proper technique block copolymers have been reported prepared by free radical polymerization.

Free Radical Polymerization at Reactive Terminus. Analogous to previously described step reaction coupling of macroreactants, Bamford has produced poly(methyl methacrylate-b-acrylonitrile) by employing the chain transfer agent, triethylamine, to cap poly-(methyl methacrylate), and then adding to this polymer free radically initiated acrylonitrile[83]. Similarly, polyvinylpyrrolidone block copolymers of methyl methacrylate[84] and acrylonitrile[85] have been prepared by cesium ammonium nitrate initiation of hydroxy terminated polyvinylpyrrolidone in the presence of methyl methacrylate or acrylonitrile. Yamashita prepared a block copolymer of styrene and methyl glutamate by terminating polystyrene with glycine, and subsequently using this polymer to initiate the ring opening polymerization of the N-carboxy anhydride (leuch's anhydride) of methyl glutamate[86].

Other workers have reported the preparation of methyl methacrylate-styrene and methyl methacrylate-vinyl chloride block copolymers by a two stage free radical polymerization. First they prepared telomers of styrene (or vinyl chloride) and poly(methyl methacrylate) terminated by carbon tetrachloride. Then they heated the telomers of the desired monomers in the presence of bis-(ephedrine) copper to prepare sequential ABAB block copolymers[87]. Poly(methyl methacrylate-b-acrylonitrile) was prepared by heating $(CO)_5Mn(CF_2)_2$-terminated poly(methyl methacrylate) and acrylonitrile at 100°C[88].

Preparation of Block Copolymers from Macroinitiators. In an early paper, Ceresa reported polymerization of methyl methacrylate in the presence of oxygen, and then swelling the polymer with styrene. On heating, the polymeric peroxide decomposed initiating styrene polymerization thus forming a block copolymer[89]. Even earlier, macroperoxides, useful in the preparation of block copolymers, were prepared by initiating the polymerization of butadiene with m-diisopropyl benzene dihydroperoxide in an emulsion. On further heating in the presence of styrene and FeII salts, a block copolymer was formed[90]. Recently in independent work, polypropylene hydroperoxide was prepared by ozonolysis, and with triethylenetetraamine as an activator was used to initiate block addition of styrene[91] and other vinyl monomers[92,93].

Block copolymers of ethylene oxide or propylene oxide and styrene were prepared by first adding the hydroxy terminated prepolymer
of the former to dodecyl isocyanate and 2,5-dimethyl-2,5 dihydro-
peroxyhexane. The resulting macroperoxycarbamate was then heated
with styrene to produce a block copolymer[94] . The late Arthur
Tobolsky prepared block copolymers of poly (ethylene glycol)
and various vinyl monomers by the preparation of a macroperoxycarbamate terminated polymer (adduct of poly(ethylene glycol) and bis (4-
isocyantocyclohexyl methane)), and then heated this prepolymer with
the desired vinyl monomer[95].

Poly(vinyl chloride-b-styrene) has been prepared by treating
-SO_3H, -CO_2H, or -CHO terminated poly(vinyl chloride) with hydrogen
chloride, and heating the resulting macrohydroperoxide with styrene[96].
In yet another variation, poly(methyl methacrylate-b-styrene) was
prepared by adding benzoyl peroxide to polystyrene-diethyl aluminum,
and subsequently heating with methyl methacrylate[97,98].

In addition to the above macroperoxides, macroazonitrile
initiators have also been employed in block copolymer preparation.
For example, poly(ethylene glycol-b-vinyl chloride) has been prepared
by capping poly(ethylene glycol) with 4,4' azobis-4-cyanovaleroyl
chloride in the presence of triethylamine. The macroazonitrile,
when heated, initiated the copolymerization of vinyl chloride,
forming a block copolymer[99]. Hydroxy terminated poly(ethylene oxide)
has been similarly capped; and also used to initiate vinyl chloride (110). An interesting block copolymer which was comprised of
methyl methacrylate and acrylamide has been prepared[100]. In the
first step chloromethylated poly(methyl methacrylate) was treated
with 2,2'-azodiisobutramidine to produce a polymeric initiator.  The
block copolymer was formed in a second step when the polymer was
heated with acrylamide.  Similarly, polystyrene macroanions were
terminated with 4,4'-azobisisobutyronitrile (Vazo), and subsequently
heated in the presence of methyl methacrylate or vinyl chloride to
produce a block copolymer[101].

Block Copolymers Prepared by Photolytic Degradation of Homopolymers.  Homopolymers with reactive end groups, such as polystyrene
terminated with bromo or tribromo groups, have been treated photo-
lytically to produce a macroradical capable of reacting with a
second monomer to produce a block copolymer[102]. Poly(methyl methacrylate-b-styrene) was prepared by irradiating polystyrene
containing about two percent copolymerized carbon monoxide in the
presence of methyl methacrylate[103]. Apparently the polystyrene
degrades at the randomly dispersed carbon monoxide mers producing
macroradicals.  These, in turn, initiate methyl methacrylate block
copolymerization.  Similarly, a styrene-carbon monoxide copolymer
in benzene was degraded in the presence of acrylic acid[104].  The
block copolymer has been shown to have interesting properties when
used as a reverse osmosis membrane.

In producing a block copolymer with interesting applications in polymer blends, Chapiro produced a block copolymer of acrylonitrile and acrylonitrile-co-styrene by gamma irradiation of the mixed monomers in dimethylformamide/benzyl alcohol[105]. It was reported that irradiation induced the formation of radical-anions. Acrylonitrile then added exclusively to the anionic chain ends, and acrylonitrile-styrene added randomly to the free radicals chain ends. The resultant polymer was found to be poly(acrylonitrile-b-acrylonitrile-co-styrene).

Preparation of Block Copolymers by Mechanical Degradation of Homopolymers. It is well known that a number of mechanical processes including mastication, milling, high speed stirring, and ultrasonic irradiation, can degrade a polymer and thus produce macroradicals. If the degradation is conducted in the presence of a second monomer block and/or graft copolymers may be formed. One of the first reported examples was the mastication of natural rubber swollen with methyl methacrylate[106]. Although small quantities of poly(methyl methacrylate) and natural rubber resulted, the yield of block copolymer was very high. Excellent discussions of the mechanical preparation of block (and/or graft) are given in several reviews[107].

O'Driscoll reports that several polymer-monomer combinations were used in the controlled preparation of homo- and block polymers by ultrasonic radiation[108]. Similarly, Fujiwara reports the preparation of poly(styrene-b-methyl methacrylate) by ultrasonic degradation of polystyrene[109]. The utility of ultrasonic irradiation is limited since yields are usually low, and the block copolymer is often contaminated with relatively large amounts of the homopolymers.

Preparation of Block Copolymers by Use of Multifunctional Initiators. The use of multifunctional initiators is a recent development in block copolymer synthesis. These initiators consist of two or more peroxide or azo moieties, each capable of monomer initiation. The initiator moieties are sufficiently (chemically) different to allow decomposition (and thus initiation) under different reaction conditions. The Pennwalt Corporation has a number of commercial and experimental difunctional initiators including di-t-butyl-4,4'-azobis-(4-cyanoperoxyvalerate).

The latter initiator was employed in the preparation of ABA poly(methyl methacrylate-b-styrene)[111]. Along with the block copolymer, a considerable amount of the respective homopolymers was also formed. Poly(methyl methacrylate-b-styrene) was prepared utilizing a sequential diperoxide initiator (Bz-OO-CO-$CH_2$-CH(Br)-CO-OO-$B_2$) containing peroxy groups of different thermal stability[112, 113]. Poly(butyl acrylate-b-methyl methacrylate) has also been prepared using a difunctional azo-initiator[114].

Russian workers have reported the preparation of "trisperoxide" or R'-OO-R-OO-R-OO-R' (where R is cyclopentylidene and R' is [(CH3)3-CO-CO+2), and used it in the preparation of poly(styrene-b-vinyl acetate)(115,116)and poly(styrene-b-vinylpyrrolidone) (117). In the latter preparation, styrene was initiated by "trisperoxide" at 75°C producing diperoxy-terminated polystyrene. This "macroinitiator" was then heated at 100°C in the presence of vinylpyrrolidone to produce an ABA copolymer.

Reactions of "Living" Macroradicals. Some polymers are insoluble in their monomers, and occluded macroradicals are produced when they are polymerized in bulk. Acrylonitrile and vinyl chloride are examples. Long lived or (psuedo-) "living" macroradicals have also been reported present when styrene is polymerized in alcohols(118), and when other vinyl monomers are polymerized in poor solvents(119). Further, it has been observed that the rate of polymerization of methyl methacrylate(120) and copolymerization of styrene-maleic anhydride(121) are faster in poor solvents than good solvents. Presumably because of the presence of long lived macroradicals. In support of these concepts Otsu reported preparing macroradicals and block copolymers of acrylamide in poor solvents for the polymer, but only homopolymer and no detectable long-lived radicals in a good solvent(129).

The above examples are all heterogeneous solution polymerizations. Heterogeneous solution polymerization occurs in most cases when the Hildebrand solubility parameter values of the solvent and polymer differ by at least 1.8H(122). The presence of macroradicals in these precipitated polymers has been demonstrated by electron spin resonance (esr)(123).

Block (or graft) copolymers have been obtained by the addition of styrene and N,N-dimethylformamide (DMF) to acrylonitrile macroradicals, and add by the addition of methyl methacrylate or styrene to vinyl chloride macroradicals(124). Block copolymers have also been produced by the addition of various vinyl monomers to styrene-maleic anhydride macroradicals (122).

When acrylic acid or vinylpyrrolidone was heated with a precipitated acrylonitrile polymer believed to contain macroradicals, block like polymers were formed(125). Likewise poly(acrylonitrile-b-methyl methacrylate) was reportedly produced when acrylonitrile was heated with precipitated methyl methacrylate polymer. Acrylonitrile-styrene copolymers however, were formed only when styrene and acrylonitrile precipitated polymer were heated in the presence of DMF. Presumably the DMF swells the polyacrylonitrile permitting diffusion of styrene monomer into the free radical containing coils. Block copolymers of methyl methacrylate and various vinyl monomers have also been reported prepared by heating monomers and methyl methacrylate macroradicals in 1-propanol(126). Interestingly poly-

(methyl methacrylate-b-styrene) was prepared by heating methyl methacrylate and styrene macroradicals in an emulsion polymerization system[127]. It has been shown that in these heterogeneous systems, the rate of copolymer formation is dependent on the initial polymerization time and initiator concentration, and requires fast (or conversely limited) decomposition, or heterogeneity of the initiator[128].

In a similar preparation styrene macroradicals have been reported when styrene was polymerized in viscous solvents and viscous poor solvents[130]. When styrene was added to these macroradicals, an increase in relative viscosity was noted. Block like copolymers were formed when either acrylonitrile, methyl methacrylate, ethyl acrylate or methacrylate was added to the macroradicals. The fastest rates of polymerization and greatest yields of copolymer were noted when polystyrene was polymerized in viscous, poor solvent[131].

It is important to note that unquestionably these workers are preparing either block, multiblock, (or possibly graft)copolymers . The yields and characterization demonstrated this fact. It is, however, not conclusive how "living" macroradicals are being produced. Similarly, it has not been shown conclusively that the primary initiator is not present (abeit at low levels). Nonetheless, copolymers of potentially interesting physical properties are being prepared, and the technique should receive consideration when evaluating block polymerization techniques.

"Living" macroradicals have been reported in other non-heterogeneous polymerization systems. For example long-lived macroradicals have been reported when methyl methacrylate was polymerized in bulk in the presence of phosphoric acid ($H_3PO_4$)[132]. Good yields of poly(methyl methacrylate-b-$\alpha$-chloroacrylate) were reported. In contrast, lower yields of block copolymer were reported when acrylonitrile was added to poly(methyl methacrylate)-$H_3PO_4$. In both cases homopolymer formation was absent.

Stable macroradicals were also reportedly prepared by the reaction of [•OH(CF$_3$)(CF$_2$)$_2$N(CF$_3$)O•] with various fluoroalkenes. They apparently reacted through the terminal nitroxide groups with fluoroalkenes, sulfur dioxide, or perfluorobutadiene to yield block copolymers of increased molecular weight[133]. An unsaturated nitrosorubber was reported when perfluorobutadiene reacted with macroradicals of perfluoropropene-1.

BIBLIOGRAPHY

1.  M. Szwarc, M. Levy, and R. Milkovich, J. Am. Chem. Soc., 78, 2656 (1956).
2.  A. Noshay and J. E. McGrath, "Block Copolymers", Academic Press, New York, 1977.
3.  R. J. Geresa (ed.), "Block and Graft Copolymerization", Vol. 1, John Wiley and Sons. New York, 1973.
4.  J. J. Burke and V. Weiss, (Eds.), "Block and Graft Copolymers", Syracuse Press, Inc., Syracuse, N.Y., 1973.
5.  D. C. Allport and W. H. Janes, (Eds.), "Block Copolymers", Applied Science Publishers, London, 1973.
6.  M. Szwarc, "Carbanions, Living Polymers, and Electron Transfer Processes", Interscience Publishers, New York, 1968.
7.  R. J. Ceresa (Ed.), "Block and Graft Copolymers", Butterworth, London, 1962.
8.  W. J. Burlant and A. S. Hoffman, "Block and Graft Polymers," Reinhold Publishing Corp., New York, 1960.
9.  Shell International Research, Brit. 1,068,130 (October 28, 1964).
10. Shell International Research, Brit. 1,098,570 (December 2, 1965).
11. S. Bywater, Adv. Polym. Sci., 4, 66 (1965).
12. M. Morton and L. J. Fetters, Macromol. Rev., 2, 71 (1967).
13. L. J. Fetters, J. Polym. Sci., C26, 1 (1969).
14. S. Horiie, et al., Jap. Kokai, 74118,689 (November 13, 1974).
15. S. Horiie, et al., Ger. Offen. 2361174 (June 12, 1975).
16. R. E. Cunningham, Ger. Offen. 2422724 (December 5, 1974).
17. M. Morton and L. J. Fetters, Rubber Rev., 48, 359 (1975).
18. L. Bi and L. J. Fetters, Macromolecules, 9, 732 (1976).
19. H. A. J. Schepers, Ger. Offen., 2442849 (March 13, 1975).
20. A. Douy, G. Jovan, and B. Gallot, C. R. Hebd. Seances Acad. Sci., Ser. C., 281 355 (1975).
21. D. J. Worsfold, Can. Pat. 973295 (August 19, 1975).
22. Y. Yamashita, K. Nobutoki, Y. Nakamuna, M. Hirota, and H. Wakabayachi, Toyoda Kenkyu Hokoku, 26, 10 (1973).
23. M. Takaki, R. Asami, and M. Ichikawa, submitted for publication in Macromolecules [see ref. 2].
24. D. H. Richards, et al., Polymer, 19, 68 (1978).
25. M. Takaki, R. Asami, H. Inukai, and T. Inenaga, Polym. Prepr., Am. Chem. Soc. Div. Polym. Chem., 18(1), 682 (1977).
26. M. Prissette, H. Magnin, M. Abadie, and F. Schue, Eur. Polym. J., 12, 713 (1976).
27. J. Baca, L. Lochmann, K. Juzl, J. Coupek, and D. Lim, J. Polym. Sci., C16, 3865 (1968).
28. D. C. Evans, J. A. Barrie, and M. H. George, Polymer, 16, 151 (1975).
29. J. M. Guyon-Gellin, J. Gole, and J. P. Pascault, J. Appl. Polym. Sci., 19, 3173 (1975).

30. J. Rossi, and B. Gallot, Makromol. Chem., 177, 2801 (1976).

31. M. Matzner, J. E. McGrath, and A. Noshay, U.S. Pat. 3,770,849 (November 6, 1973).

32. Y. Yamashita, M. Matsui, and K. Ito, J. Polym. Sci., Polymer Chem. Ed., 10, 3577 (1972).

33. J. Stehlicek and J. Sebenda, Polyamidy '75, 91 (1975).

34. H. Kawamoto, Jap. Kokai 74 30 707 (August 15, 1974).

35. S. A. Sundet, et al., Macromolecules, 9, 371 (1976).

36. R. P. Foss, H. W. Jacobson, H. N. Cripps, and W. H. Sharkey, Macromolecules, 9, 373 (1976).

37. D. R. Jackson and L. G. Lundsted, U.S. Pat. 2,677,700 (1974).

38. D. R. Jackson and L. G. Lundsted, U.S. Pat. 3,036,130 (1962).

39. D. H. Richards and M. Szwarc, Trans. Fara. Soc., 55, 1644 (1959).

40. G. Finaz, Y. Gallot, J. Parod, and R. Rempp, J. Polym. Sci., 58, 1363 (1962).

41. J. J. O'Malley and R. H. Marchessault, Macromol. Synth., 3, 35 (1972).

42. S. Marti, J. Nervo, and G. Riess, Progr. Colloid and Polymer. Sci., 58, 114 (1975).

43. G. Riess, J. Nervo, and D. Rogez, Polym. Prep., Am. Chem. Soc. Div. Polym. Chem., 18(1), 329 (1977).

44. Y. Minoura and A. Nakano, Macromol. Synth., 4, 25 (1972).

45. M. Tomoi, Y. Shibayama, and H. Kakiuchi, Polym. J., 8, 190 (1976).

46. C. W. Brown and I. F. White, J. Applied Polym. Sci., 16, 2671 (1972).

47. P. T. Hale and G. A. Pope, Eur. Polym. J., 11, 677 (1975).

48. A. Douy, G. Jouan, and B. Gallot, Makromol. Chem., 177, 2945 (1976).

49. P. K. Wong, A. E. Zachariades, and M. Szwarc, Polymer, 17, 817 (1976).

50. C. U. Pittman, Jr. and A. Hirao, J. Polym. Sci., Polym. Chem. Ed., 15, 1677 (1977).

51. J. P. Kennedy and E. G. Melby, J. Polym. Sci., Polym. Chem. Ed., 13, 29 (1975).

52. J. P. Kennedy, E. G. Melby and A. Vidal, J. Macromol. Sci., A9, 833 (1975).

53. J. P. Kennedy and A. Vidal, J. Polym. Sci., Polym. Chem. Ed., 13, 1765 (1975).

54. J. M. Rooney, D. R. Squire, and V. T. Stannett, J. Polym. Sci., Polym. Chem. Ed., 14, 1877 (1976).

55. R. J. Perry, U.S. Pat. 3,859,387 (January 7, 1975).

56. M. Koga and H. Sato, Ger. Offen. 2,508,146 (August 28, 1975).

57. V. S. Shteinbak, V. V. Amerik, F. I. Yakobson, Y. S. Kissin, D. V. Ivanyukov, and B. A. Krentsel, Eur. Poly. J., 11, 457 (1975).

58. Y. Takeshita and Y. Shuts, Jap. Kokai 75, 141, 694 (November 14, 1975), C. A. 84, 151263y (1976).

59.  G. A. Clegg, D. R. Gee, and T. P. Melia, Makromol. Chem.,
     132, 203 (1970).
60.  H. W. Coover, Jr., R. L. McConnel, F. B. Joyner, D. F. Slonaker,
     and R. L. Combs, J. Polym. Sci., A1, 4, 2563 (1966).
61.  A. W. Langer, U.S. Pat. 3, 450, 795 (1969).
62.  H. Tanzawa, T. Tanaka, and A. Soda, J. Polym. Sci., A2, 7,
     929 (1969).
63.  V. V. Americk, D. V. Inayvkov, F. I. Yakobson, and B. A.
     Krentsel, Vysokomol. Soedin., B15, 500 (1973); CA80, 37481k
     (1974).
64.  E. Agouri, C. Parlant, P. Mornet, J. Rideau, and J. F. Teitgen,
     Makromol. Chem., 137, 229 (1970).
65.  N. S. Chu and J. L. Jezl, Fr. Pat. 1, 531, 409 (July 5, 1968).
66.  W. J. S. Craven, S. African Pat. 67/06, 649 (December 8, 1966).
67.  P. Teyssie, et al., "Ring Opening Polymerization," ACS Symp.
     Ser., 59, 165 (1977).
68.  J. Herman, R. Jerome, P. Teyssie, M. Gervais, B. Gallot,
     J. Makromol. Chem., 179, 1111 (1978).
69.  Y. Iwakura, K. Uno, T. Kusunoshita, and T. Nakanishi, Jap.
     Kokai 74 25, 439 (June 29, 1974).
70.  B. M. Baysal, E. H. Orhan, and I. Yilgor, Proje Kesin Rap. -
     Turk. Bilimsel Tek. Arastirma Kurumu. Polim. Kim. Unitesi.,
     11, 31 (1974); CA, 83, 179667 m.
71.  L. Pohjola, O. Harva, and J. Karvinen, Finn. Chem. Lett., 1974,
     p 221.
72.  W. L. Hergenrother and R. J. Ambrose, J. Polym. Sci., Polym.
     Lett. Ed., 12, 343 (1974).
73.  W. L. Hergenrother and R. J. Ambrose, J. Polym. Sci., Polym.
     Chem. Ed., 12, 2613 (1974).
74.  R. J. Ambrose and W. L. Hergenrother, J. Polym. Sci., Polym.
     Symp., 60, 15 (1977).
75.  J. H. Grezlak and G. L. Wilkes, J. Appl. Polym. Sci., 19, 769
     (1975).
76.  Y. Yamashita, N. Suzuki, and K. Nobutoki, Kobunshi Ronbunshu.,
     31, 623 (1974); CA, 82, 112279x.
77.  D. M. White, U.S. Pat. 3, 875, 256 (April 1, 1975).
78.  C. Minazzi, J. Esnault, and A. Pleurdeau, Makromol. Chem.,
     177, 663 (1976).
79.  R. J. Zdrahala, E. M. Firer, and J. F. Fellers, J. Polym. Sci.,
     Polym. Chem. Ed., 15, 689 (1977).
80.  H. R. Musser and W. J. Jackson, Jr., U.S. Pat. 3, 943, 189
     (March 9, 1976).
81.  M. Sumoto, K. Tsuji, H. Furusawa, Jap. Kokai 74 48, 195
     (December 19, 1974).
82.  T. Okano, M. Ikemi, and I. Shinohara, Polymer J., 10, 477
     (1978).
83.  C. H. Bamford and E. F. T. White, Trans. Fara. Soc., 54, 268
     (1958).
84.  L. J. Guilbault, J. Macromol. Sci., Chem. Ed., A7, 1581 (1973).

85. L. J. Guilbault, U.S. Pat. 3, 907, 927 (September 23, 1975).
86. Y. Yamashita, Y. Iwaya, and K. Ito, Makromol. Chem., 176, 1207 (1975).
87. A. Guyot, M . Ceysson, A. Michel, and A. Revillon, Inf. Chim., 116, 127 (1973).
88. C. H. Bamford and S. U. Mullik, Polymer, 17, 94 (1976).
89. R. J. Ceresa, Polymer, 1, 397 (1960).
90. R. J. Orr and H. L. Williams, J. Am. Chem. Soc., 79, 3137 (1957).
91. E. Bigdeli, R. W. Lenz, B. Oster, and R. D. Lundberg, J. Polym. Sci., Polym. Chem. Ed., 16, 469 (1978).
92. G. Smets, G. Weinand, and S. Deguchi, J. Polym. Sci., Chem. Ed., 16, 3077 (1978).
93. G. Weinand and G. Smets, J. Polym. Sci., Chem. Ed., 16, 3091 (1978).
94. B. M. Baysal, E. H. Orhan, and I. Yilgor, J. Polym. Sci., Polym. Symp., 46, 237 (1973).
95. A. V. Tobolsky, U.S. Pat. 3, 865, 898 (February 11, 1975).
96. P. Lalet, H. Fassy, and A. Miletto, Fr. Pat. 2, 213, 293, (August 2, 1974).
97. E. B. Milovskaya and L. V. Zamoiskaya, Vysokomol. Soedin., Ser. B, 18, 300 (1976).
98. L. V. Zamoiskaya, E. S. Gankina, and E. B. Milovskaya, Vysokomol. Soedin., Ser A, 18, 1635 (1976).
99. J. J. Laverty and Z. G. Gardlund, Polym. Prepr., Am. Chem. Soc. Div. Polym. Chem., 15 (2), 306 (1974).
100. S. N. Gupta and U. Nandi, Makromol. Chem., 176, 3179 (1975).
101. Y. Vinchon, R. Reeb, and G. Riess, Eur. Polym. J., 12, 317 (1976).
102. A. S. Dunn, B. D. Stead, and H. W. Melville, Trans. Fara. Soc., 50, 279 (1954).
103. W. Kawai, J. Polym. Sci., Chem. Ed., 15, 1479 (1977).
104. W. Kawai, J. Macrolmol. Sci., Chem. Ed., A11, 1027 (1977).
105. A. Chapiro, F. Jaberg, and L. Perec-Spritzer, Eur. Polym. J., 11, 637 (1975).
106. D. J. Angier and W. F. Watson, J. Polym. Sci., 25, 1 (1957).
107. A. Casale and R. S. Porter, Adv. Polym. Sci., 17, 1 (1975).
108. K. F. O'Driscoll and A. U. Sridharan, Appl. Polym. Symp., 26, 135 (1975).
109. H. Fujiwara, H. Kakiuchi, K. Katsuhiko, and K. Goto, Kobunshi Ronbunshu, 33, 183 (1976).
110. J. J. Laverty and Z. G. Gardlund, J. Polym. Sci., Polym. Chem. Ed., 15, 2001 (1977).
111. I. Piirma and H. Chou, Polym. Prepr., ACS Div. Polym. Chem., 16 (2), 506 (1975).
112. N. G. Ivanova, S. S. Ivanchev, and N. M. Domareva, Vysokomol. Soedin., Ser. A., 18, 2788 (1976).
113. S. S. Ivanchev, N. G. Ivanova, Y. L. Zherebin, Plaste Kautsch, 23, 5 (1976).

114.  I. Piirma and B. Gunesin, Polym. Prepr., ACS Div. Polym. Chem.,
      18 (1), 687 (1977).
115.  A. I. Kirillov, A. D. Molinkov, and Y. A. Zvereva, Russ. 440,
      378 (August 25, 1974).
116.  G. D. Rudkovskaya, I. N. Nikonova, I. A. Baranovskaya,
      V. A. Belov, and T. A. Sokolova, Vysokomol. Soedin., Ser. A,
      17, 13 (1975); CA, 82, 140691u.
117.  G. D. Rudkovskaya, I. A. Baranovskyaya, I. N. Nikonova,
      T. A. Komogorova, and T. A. Sokolova, Vysokomol. Soedin.,
      Ser. B, 17, 786 (1975); CA, 84, 44734f.
118.  A. J. Chapiro, Chem. Phys., 47, 747 (1950).
119.  J. Hiemeleers and G. Smets, Makromol. Chem., 47, 7 (1961).
120.  B. Atkinson and G. R. Cotten, Trans. Fara. Soc., 54, 877
      (1958).
121.  E. U. Tsuchida, et al., J. Chem. Soc. Japan, 70, 573 (1967).
122.  R. B. Seymour, H. S. Tsang, E. E. Jones, P. D. Kincaid, and
      A. K. Patel, Advan. Chem. Ser., 99, 418 (1971).
123.  C. H. Bamford, J. Polym. Sci., 48, 37 (1960).
124.  Y. Minoura and Y. Ogota, J. Polym. Sci., A-1, 7, 2547 (1969).
125.  R. B. Seymour, D. R. Owen, G. A. Stahl, H. Wood, and W. N. Tin-
      nerman, Appl. Polym. Symp., 25, 69 (1974).
126.  R. B. Seymour, G. A. Stahl, D. R. Owen, and H. Wood, Advan.
      Chem. Ser., 142, 309 (1975).
127.  K. Horie and D. Mikulasova, Makromol. Chem., 175, 2091 (1974).
128.  D. Mikulasova, V. Chrastova, P. Citovicky, and K. Horie,
      Makromol. Chem., 178, 429 (1977).
129.  H. Tanaka, T. Sato, and T. Otsu, Macromol. Chem., 180, 267
      (1979).
130.  R. B. Seymour, G. A. Stahl, and H. Wood, Appl. Polym. Symp.,
      26, 249 (1975).
131.  G. A. Stahl, R. B. Seymour, and D. P. Garner, Coatings and
      Plastics Prepr., ACS Div. of Organic Coatings and Plastics
      Chemistry, 37 (1), 714 (1977).
132.  A. V. Olevin, M. B. Lachinov, V. P. Zubov, and V. A. Kobonov,
      Vysokomol. Soedin., Ser. B, 18, 219 (1967); CA, 85, 6172j.
133.  A. Arfaei, R. N. Haszeldine, and S. Smith, J. Chem. Soc., Chem.
      Commun., 1976, 260.

# PREPARATION OF BLOCK COPOLYMERS FROM STYRENE-ACRYLONITRILE

# MACRORADICALS

Raymond B. Seymour, David P. Garner,* G. Allan Stahl,**
Roger D. Knapp,† and Laura Sanders††

Department of Polymer Science
University of Southern Mississippi
Hattiesburg, MS  39401

## ABSTRACT

Charge transfer complexes of styrene and acrylonitrile have
been shown to exist when in the presence of zinc chloride.  Proton
nuclear magnetic resonance spectroscopy has been used to establish
this effect.  In the proper solvents styrene and acrylonitrile will
form occluded macroradicals which may then be used to form block
copolymers.  These block copolymers occur both in the presence and
absence of zinc chloride.  Pyrolysis gas chromatography, differential scanning calorimetry, and solubility studies show the properties of the two copolymers and their various block copolymers to
be quite similar.  Differences in the copolymers may be seen from
carbon-13 nuclear magnetic resonance spectroscopy.  Yield data for
the block copolymers is reported.

## INTRODUCTION

The high level of interest in block copolymers in the academic
and industrial communities is demonstrated by its fast growing
literature.  Any number of quality discussions of block copolymer
preparation and properties are available.  Among the better are
the predecessor to this volume (1), and the reviews of Noshay and
McGrath (2), Ceresa (3), Burke and Weiss (4), and Allport and

*General Motors Research Laboratories, Warren, MI  48090
**B.F. Goodrich Research and Development Center, Brecksville,
  OH  44141
†Department of Biochemistry, Baylor College of Medicine,
  Houston, TX  77025
††Lone Star Industries, Houston, TX  77032

Janes (5). We have concentrated our efforts on the preparation of these materials and have attempted to prepare them by inexpensive yet effective techniques. Free radical polymerization was selected for study in this work over other alternatives, such as reactions of macro-anions, coupling of polymer termini, and the use of coordination catalysts.

In past studies, block copolymers have been prepared from free radicals by several techniques including macroradical coupling, photolytic or mechanical degradation in the presence of a second monomer, the use of difunctional initiators, and reactions of long lived macroradicals. Poly(methyl methacrylate-b-acrylonitrile), for example, has been prepared by employing the chain transfer agent, triethylamine, to cap poly(methyl methacrylate) and then adding this polymer to free radically initiated acrylonitrile (6). Methyl methacrylate-vinyl chloride block copolymers were prepared by coupling carbon tetrachloride terminated poly(methyl methacry- late) and vinyl chloride telomers (7). The coupling agent was bis- (ephedrine) copper. A macroinitiator was similarly produced by treating chloromethylated poly(methyl methacrylate) with 4,4'-azodi- isobutramidine (8). This polymeric initiator was then heated in the presence of acrylamide to produce a hydrophilic block copolymer.

Homolytic bond cleavage, producing free radicals used in block copolymer formation, has also been reported. One of the first examples was the mastication of natural rubber swollen with methyl methacrylate (9). Photolytic degradation of polystyrene, capped with bromine containing groups (10) or copolymerized with carbon monoxide (11), has also been used in block copolymer preparation. Similarly ultrasonic degradation of homopolymers in the presence of a second monomer has been reported (12).

A more recent development in block copolymer preparation is the use of difunctional initiators. These initiators, usually diperoxides or peroxide-azo compounds, have been used to prepare block copolymers of vinyl acetate (13) and styrene (14,15). So far the technique has not received a great deal of acceptance because the product is heterodispersed and contains a great deal of homo- polymer.

Block copolymers have also been produced by the addition of vinyl monomers to occluded or long lived macroradicals. These "pseudo" living macroradicals are produced when vinyl monomers are polymerized in poor solvents (16-18) or in viscous medium (19). These occluded macroradicals have been used to prepare block copoly- mers of styrene (17,19), acrylonitrile (20), vinyl acetate (21), and methyl methacrylate (22). This principal may be extended to binary monomer systems as well. An interesting example of this is shown by the highly alternating copolymers synthesized by charge transfer complex (CTC) copolymerization.

Monomer pairs, one electron rich, the other electron poor, have been shown to form (CTC)'s. Typical CTC's are formed from styrene and maleic anhydride and styrene and acrylonitrile in the presence of Lewis acids (23-25). These CTC's are known to rapidly polymerize in the presence of free radical initiators to form copolymers with a high degree of alternation (26,27). If the copolymerization is conducted in a poor solvent for the polymer, occluded macroradicals will be produced.

In this paper we will discuss evidence for the existence of CTC's of styrene and acrylonitrile and the resultant differences in the copolymer produced from them. We will also report the preparation of copolymers with block copolymer characteristics from poly(styrene-co-acrylonitrile) macroradicals prepared in tert-butyl alcohol. This solvent is a poor solvent for the macroradical and exhibits a very low chain transfer constant (28). The copolymers reported in this paper were prepared in the absence and presence of zinc chloride (ZnCl$_2$), and the effect of ZnCl$_2$ on the reactivity of the macroradical will be discussed.

The styrene-acrylonitrile copolymers and block copolymers were characterized by selective solvent fractionation, $^1$H NMR, pyrolysis gas chromatography, and differential scanning calorimetry.

EXPERIMENTAL

Prior to use, the monomers, styrene (S), acrylonitrile (AN), methyl methacrylate (MMA), and vinylpyrrolidone (VP) were vacuum distilled and stored in brown bottles at 0°C. The technical grade solvents were used as received.

A typical polymerization consisted of equimolar amounts of styrene (S) and acrylonitrile (AN (0.02 moles). They were co-polymerized with and without 10% mole ratio ZnCl$_2$ (0.002 moles) in tert.-butyl alcohol (20 ml) in the presence of the initiator tert.-butyl peroxypivalate (tBPP, Lupersol-11, 0.05 g). The polymerizations were conducted at 50°C in 1 oz. amber bottles for 96 hrs. After 96 hrs., either methyl methacrylate (MMA), acrylonitrile (AN), styrene (S), or vinylpyrrolidone (VP)(1 ml) was added to the precipitated SAN macroradicals, and allowed to react 96 additional hours. Other S-AN-ZnCl$_2$ copolymerizations in the respective molar ratios of 5:10:1 (0.02:0.04:0.004 moles) were carried out in tert.-butyl alcohol (200 ml) with tBPP (0.15 g). The polymerizations were carried out at 50°C and 70°C in amber bottles for 4 hrs.

A Varian T-60 $^1$H, NMR, and Carey-14 UV spectrometers were employed to detect the presence of the CTC. High resolution $^1$H noise decoupled $^{13}$C NMR spectra were obtained by use of a Varian XL-100-15 Spectrometer, and Nicolet Technology TT-100 Data System. All spectra were taken in deuteroacetone.

Pyrolysis gas chromatograms (PGC) were obtained on a model AIOOC Wilkens Aerograph gas chromatograph equipped with a Leeds and Northrup Speedomax G Recorder. The copper chromatography column was packed with acid washed chromosorb W and 20% SE-20. Pyrolysis was accomplished by placing a small, dry sample of polymer on a Rh-W, code 13-002, Gow-Mac coil and pyrolyzing for a predetermined time.

A DuPont Model 990 thermal analyzer equipped with a DSC cell was used for differential scanning calorimetry. All measurements were made in a nitrogen atmosphere over a temperature range from room temperature to 150°C.

Polymerization rates were determined by recording capacitance dilatometry[29]. In these experiments S (1.90 g), AN (1.1 g), and the appropriate solvent (50 ml) were heated in the dilatometer with tBPP (0.06 g) and $ZnCl_2$ (0.22 g) when required.

## RESULTS AND DISCUSSION

Evidence for the CTC system of S and An-$ZnCl_2$, that we have chosen to work with, was demonstrated by $^1H$ NMR. The center of the AN vinylic proton absorbance is at 5.90 δ. When $ZnCl_2$ is added in a 0.10 mole ratio, the center of the absorbance shifts downfield to 6.00 δ. This corresponds to removal of electrons from AN, making the monomer more electron deficient. Upon addition of S (S:AN:$ZnCl_2$, 10:10:1) the center of the AN absorbance shifts to 5.60 δ. This corresponds to a donation of electrons from S to the AN-$ZnCl_2$ complex resulting in a three-membered complex with electron density flow given by S→AN→$ZnCl_2$. The S vinylic peaks are observed to shift downfield approximately 0.10 δ each. Figure 1a shows the $^1H$ NMR spectra for AN, and Figure 1b shows the effect of $ZnCl_2$ on AN at a 10:1 ratio. Figure 1c shows the $^1H$ NMR of the CTC of S, AN, and $ZnCl_2$. Copolymers of S and AN formed in the presence of $ZnCl_2$ should exhibit a more highly alternating mer sequence than copolymers formed in the absence of $ZnCl_2$.

As shown in Figure 2, the rate of SAN copolymerization in the presence of $ZnCl_2$ is solvent dependent. The greatest rates of polymerization were noted in 1-pentanol (solubility parameter δ=10.9 H) and tert.-butyl alcohol (δ=10.6 H). The rates were correspondingly slower in less polar 1-octanol (δ=10.3 H), and in more polar 1-butanol (δ=11.4 H) and 1-propanol (δ=11.9 H). PSAN swells slightly in 1-octanol, but is insoluble in the other alcohols. Presumably the optimum solubility parameter range 10.6-10.9 H (the solubility parameters of 1-pentanol and tert.-butyl alcohol) corresponds to the optimum solvency of the CTC and the optimum minimum solvation of the copolymer. The slower rates in 1-butanol and 1-propanol, both poor solvents for the copolymer, are probably due to the inability of the polarized CTC to penetrate the non-polar macroradical coil. The slower rate of

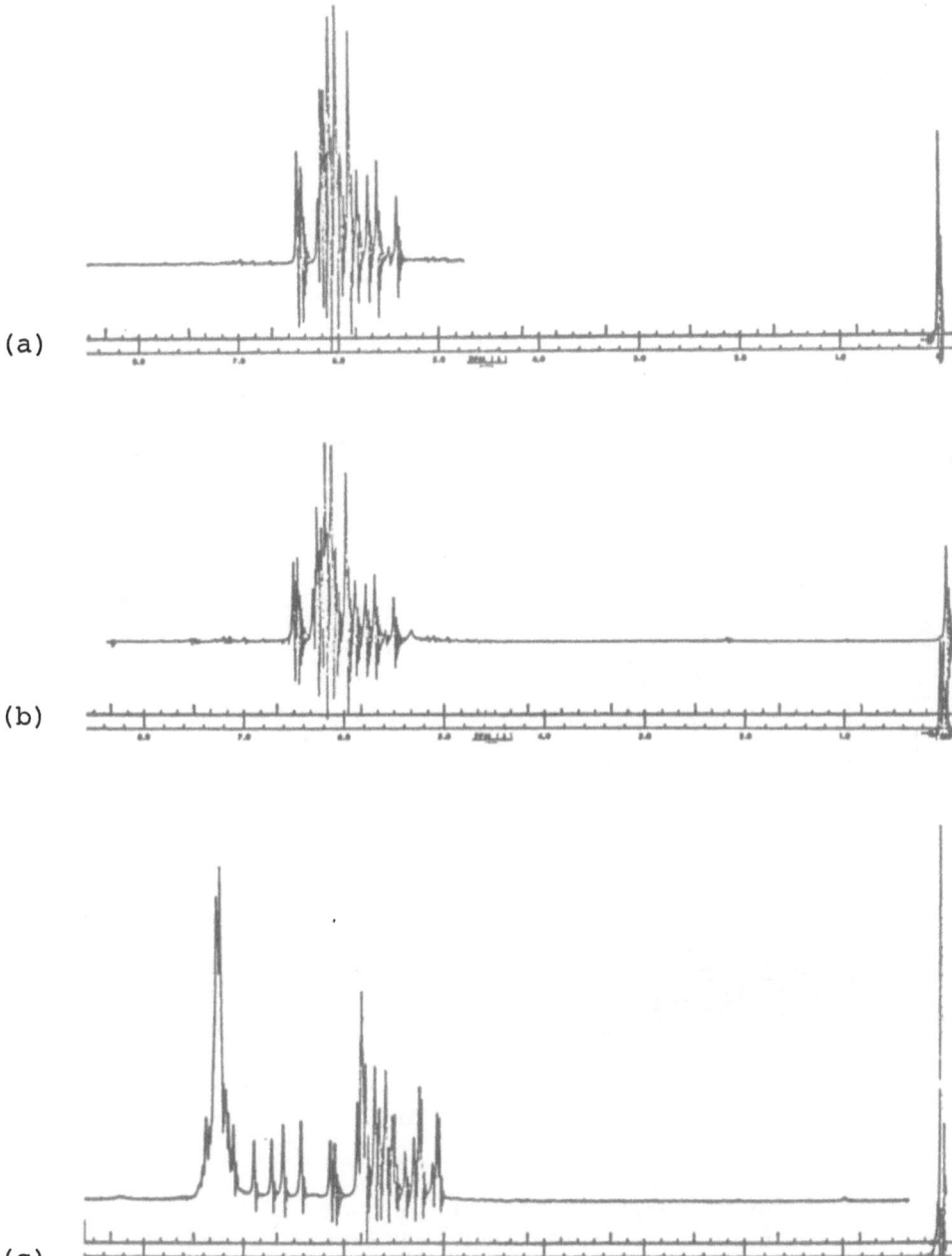

Fig. 1.  ¹H NMR spectra for (a) acrylonitrile (b) acrylonitrile
complexed with ZnCl₂ in a 10:1 ratio (c) styrene, acrylonitrile, ZnCl₂ complexed in a 10:10:1 ratio.

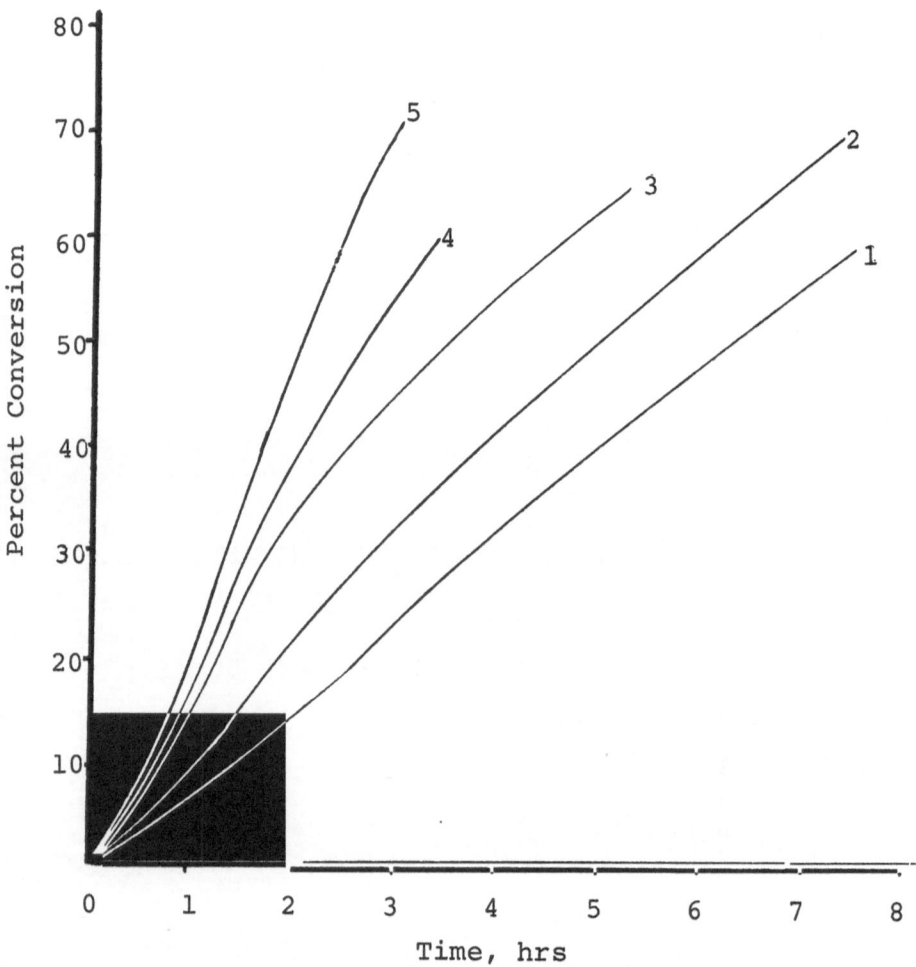

Fig. 2.   Rates of polymerization of S and AN (1:1) with ZnCl$_2$ at
          50°C in 1) 1-propanol, 2) 1-butanol, 3) 1-octanol, 4)
          tert-butyl alcohol, and 5) 1-pentanol.

copolymerization in 1-octanol is not surprizing since 1-octanol
swells the polymer thus promoting increased macroradical termination.

The rates of copolymerization of SAN in the absence of $ZnCl_2$
were measured in tert.-butyl alcohol and 1-octanol.  The rates,
shown in Figure 3, are slower than the corresponding rates with
$ZnCl_2$ present.  It has been suggested by Gaylord that the rates
of copolymerization are enhanced by the presence of CTC's which
exist as "organized mobile monomer arrays (30).  Simply stated,
without $ZnCl_2$ the monomers exist in solution without order, and
thus add more slowly and in a random fashion.

The reactivity ratios for S and AN in the S/AN 1:1 system are
0.40 and 0.04, respectively (31).  A strict alternation of units
is not followed, but for equimolar copolymerization, a certain
degree of alternation is observed.  Thus, the copolymers produced
at 50°C with and without $ZnCl_2$ should have the greatest differences
in composition.  Yet, a comparison of PGC pyrograms for these copolymers showed only a 10% difference in the characteristic peak
areas.  However, the mer arrangements are quite different as shown
by Figures 4a and 4b.  The $^{13}C$ NMR spectrum of the nitrile carbon
shown in Figure 4a is of the copolymer prepared in the absence of
$ZnCl_2$.  Note that Figure 4a displays one peak which corresponds
to the -S-AN-S- sequence, while Figure 4b, the nitrile for the
copolymer prepared in the absence of $ZnCl_2$, shows two peaks.  The
smaller peak is due to the -S-AN-AN- mer sequence.  Thus while
the alternation of mers in both copolymers is very likely, as shown
by its being the major peak, it is almost the only arrangement in
the copolymer prepared with $ZnCl_2$.

Despite the differences in the mer sequencing, physical properties of the two copolymers are very similar.  As mentioned, the
PGC pyrograms show only small changes.  The DSC thermograms show
smaller differences with the Tg's for the copolymers being about
the same.  Furthermore, the solubilities of the two copolymers are
the same, and our fractionation scheme for both copolymers and their
respective block copolymers is identical.

Increased rates of polymerization have classically been
explained by a decrease in the rate of termination.  We can take
advantage of the resulting longer-lived polymeric radicals by
proper selection of conditions and careful experimental techniques
to avoid contamination with possible termination agents such as
oxygen.  Consider the copolymerization of SAN in tert.-butyl
alcohol.  The resulting copolymer precipitates, and is capable of
further polymerization even after 96 hrs.  The latter interval
was selected to allow near quantitative decomposition of the
primary initiator.  Representative of the extended reactivity,
when MMA was added to precipitated PSAN, a copolymer with block

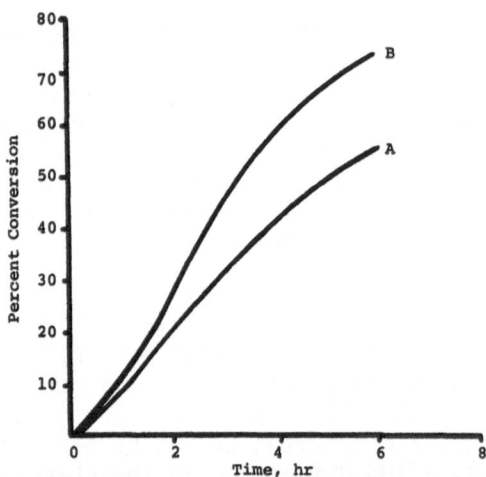

Fig. 3.   Rates of polymerization of S and AN (1:1) in the absence
          of ZnCl$_2$ at 50°C in A) 1-octanol and B) tert-butyl alcohol.

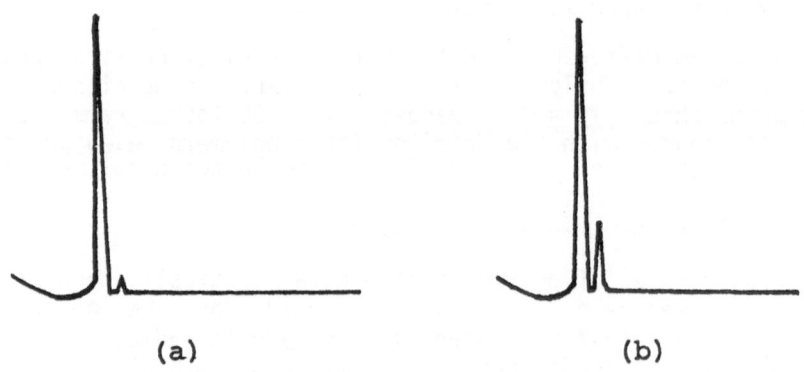

Fig. 4.   $^{13}$C NMR of the nitrile carbon of styrene/acrylonitrile
          copolymers prepared in the (a) presence of ZnCl$_2$ and
          (b) in the absence of ZnCl$_2$.

copolymer characteristics was produced. The results of these
experiments are given in Table I. As shown, an increase in weight
of 99.0% was observed when MMA was added to precipitated PSAN in
the absence of $ZnCl_2$. Solvent fractionation indicated the increase
corresponds to 67.9% PSAN-b-MMA, and 18.6% and 13.9% of SAN and MMA
homopolymers. When $ZnCl_2$ was present, a weight increase of 96.8%
was observed corresponding to 54.4% block copolymer, and 35.6 and
9.8% SAN and MMA homopolymers. After 96 hrs, the initiator tBPP,

Table I. Composition of Polymeric Products Obtained when MMA and
SAN Macroradicals were Heated for 96 hours at 50°C

| $ZnCl_2$ | % Poly(SAN) | % Poly(MMA) | % Poly(SAN-b-MMA) | % Total Yield |
|---|---|---|---|---|
| Not present | 18.6 | 13.4 | 67.9 | 99.0 |
| Present | 35.6 | 9.8 | 54.5 | 96.8 |

had gone through several half-lives, and should be non-existent.
We must, therefore, assume that the polymerization of MMA is
initiated by PSAN macroradicals. Figure 5 shows a DSC scan of
poly(S/AN-b-MMA). (DSC scans for S/AN made with and without $ZnCl_2$
showed little difference in Tg's for the copolymer or the blocks.)

Similarly, when S is added to precipitated PSAN in the absence
and presence of $ZnCl_2$, block copolymers were produced in 69.8% and
42.7% yields. These results are tabulated in Table II.

Table II. Composition of Polymeric Products Obtained when S and
SAN Macroradicals were Heated for 96 hours at 50°C

| $ZnCl_2$ | % Poly(SAN) | % Poly(S) | % Poly(SAN-b-S) | % Total Yield |
|---|---|---|---|---|
| Not present | 30.2 | trace | 69.8 | 84.8 |
| Present | 48.8 | 8.3 | 42.7 | 77.9 |

Interestingly in the absence of $ZnCl_2$, only traces of PS were
formed while 8.3% PS was produced when $ZnCl_2$ was present. Also
the total yield of polymer (PSAN + PS + PSAN-B-S) is 84.8% and
77.9% without and with $ZnCl_2$. A DSC scan for a poly(S/AN-b-S)
copolymer is shown in Figure 6.

When AN was added to SAN macroradicals in the absence and
presence of $ZnCl_2$, 49.6% and 41.1% block copolymers were produced,

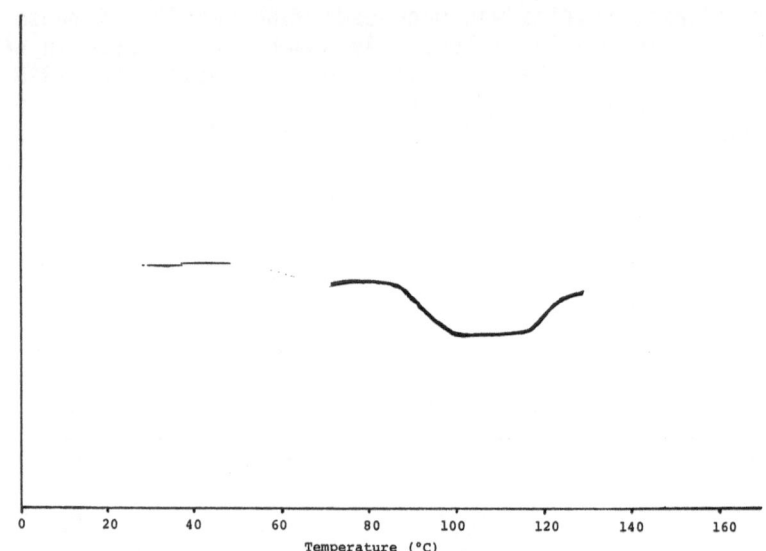

Fig. 5.  DSC thermogram of styrene/acrylonitrile-b-methyl
         methacrylate copolymer.

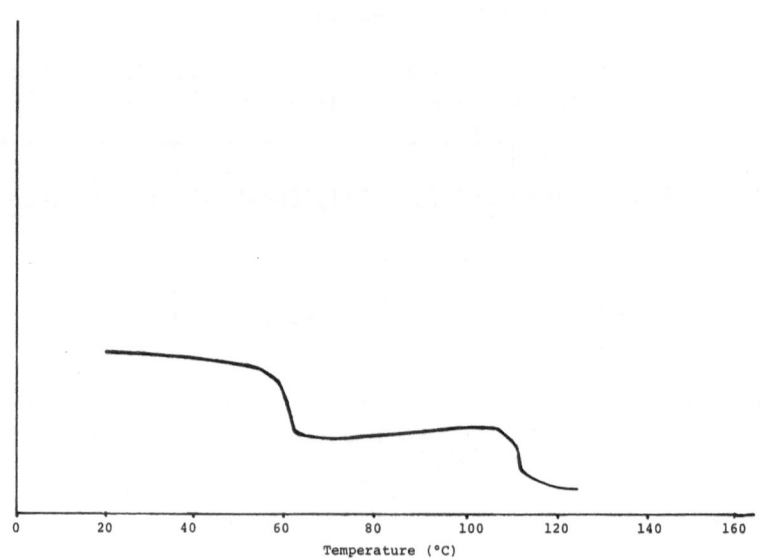

Fig. 6.  DSC thermogram of styrene/acrylonitrile-b-styrene copolymer

respectively.  These and related data are presented in Table III.
Figure 7 shows a DSC scan for poly(S/AN-b-AN).

Table III. Composition of Polymeric Products Obtained When AN and
        SAN Macroradicals were Heated for 96 hours at 50°C

| ZnCl$_2$ | % Poly(SAN) | % Poly(AN) | % Poly(SAN-b-AN) | % Total Yield |
|---|---|---|---|---|
| Not present | 32.3 | 18.1 | 49.6 | 99.0 |
| Present | 45.1 | 13.4 | 41.4 | 99.0 |

As shown in Table IV, the total yield of polymer products produced is decreased when VP is added to SAN macroradicals.  There
was 85.4% total yield in the absence of ZnCl$_2$ and 70.8% total yield
in the presence of ZnCl$_2$.  This probably corresponds to a higher
rate of chain transfer of VP in radical polymerization.  The yields
of the corresponding block copolymers were, in the absence of ZnCl$_2$,
64.9%, while in the presence of ZnCl$_2$, 60.0%.

Table IV. Composition of Polymeric Products Obtained when VP and
        SAN Macroradicals were Heated for 96 hours at 50°C

| ZnCl$_2$ | % Poly(SAN) | % Poly(VP) | % Poly(SAN-B-VP) | % Total Yield |
|---|---|---|---|---|
| Not present | 35.1 | trace | 64.9 | 85.4 |
| Present | 40.0 | trace | 60.0 | 70.8 |

The block copolymers reported above were characterized by PGC
and DSC as well as solvent fractionation.  The existence of both
monomers in the polymer was shown by PGC.  DSC thermograms showed
two distinct regions corresponding to block-like addition rather
than random addition.

SOLVENT FRACTIONATION

The block copolymers obtained from the reaction of SAN macroradicals were purified by solvent fractionation.  In general the
gross polymer was dissolved in a suitable solvent and then frac-
tionally precipitated by adding a poor solvent.

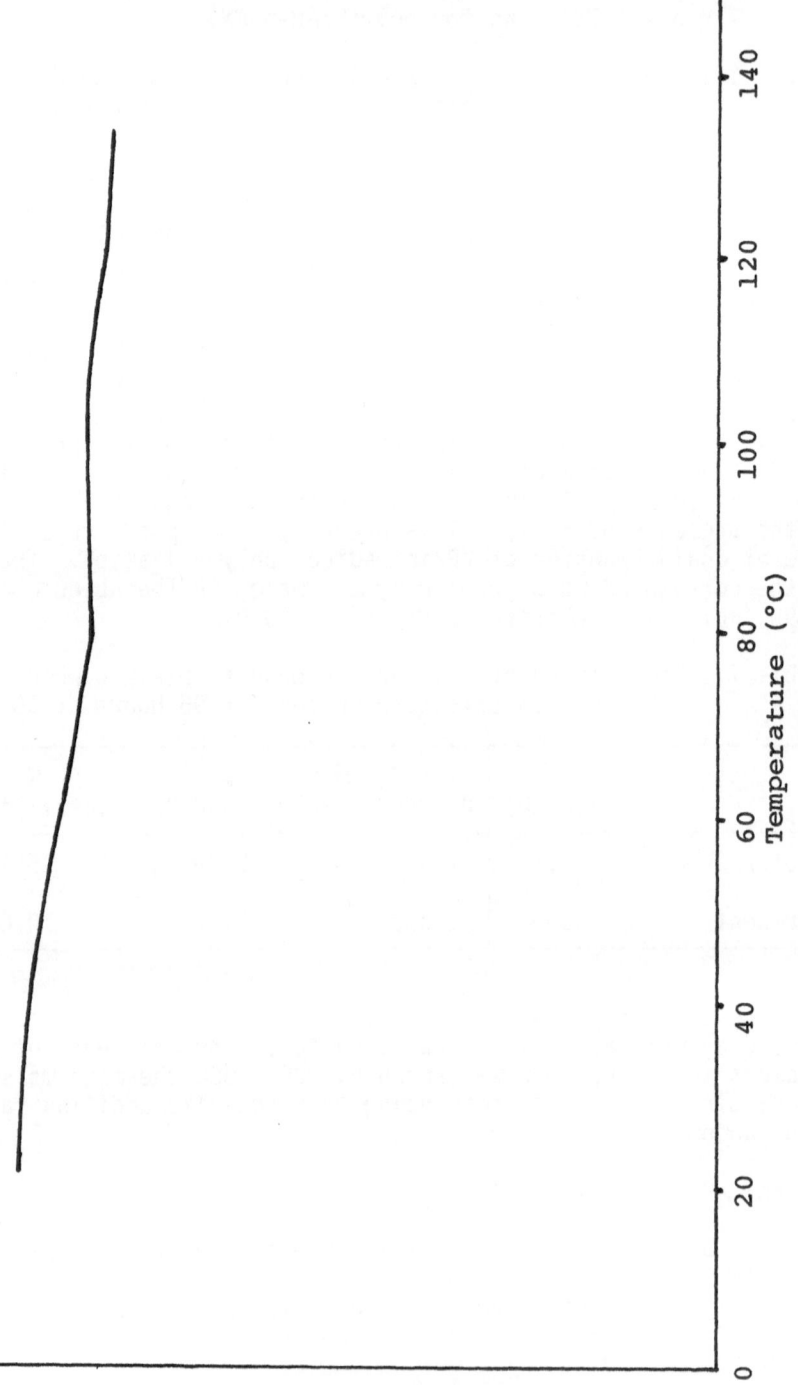

Fig. 7. DSC thermogram of styrene/acrylonitrile-b-acrylonitrile copolymer.

Copolymers of SAN and S or MMA were purified by dissolving in dimethylformamide.  PS or PMMA was precipitated by the addition of 0.8 volume parts of methanol.  The block copolymer was then removed by increasing the methanol to 1.5 volume parts.  The remaining PSAN was then precipitated by increasing the methanol to 2.2 volume parts.

PSAN-b-AN was purified by extracting with acetone to remove PSAN and some PSAN-b-AN.  Addition of 0.5 volume parts petroleum ether precipitated the latter from the extract.  The non-acetone soluble material was then dissolved in dimethylformamide and PSAN-b-AN removed by addition of up to 2.2 volume parts methanol.

PSAN-b-VP was purified by first dissolving the gross polymer in acetone and sequentially adding methanol.  PSAN precipitated on the addition of 0.2-1.0 volume parts methanol, PSAN-b-VP with 1.0-2.0 volume parts, and PVP with greater than 2.0 volume parts.

The composition of the individual fractions were confirmed by PGC.

CONCLUSIONS

This work demonstrates that the technique of occluded macro-radicals may be used to produce block copolymers from highly alternating copolymer systems.  However, the following points must be mentioned.  (1) In all cases when $ZnCl_2$ was present, the yields of termination S/AN copolymers were higher with a corresponding decrease in the yields of the S/AN block copolymers.  The overall yields showed some change as well, but generally not as great as for the previously mentioned yields.  We therefore feel that the $ZnCl_2$ or some impurity in the $ZnCl_2$ is acting as a chain transfer agent.  (2) Physical properties of both types of copolymers and block copolymers are virtually the same.

ACKNOWLEDGEMENTS

The authors would like to thank Ernest Massatenta and Jane Fiebelkorn for their assistance in the preparation of this manuscript.

REFERENCES

1.  D. Klempner and K. C. Frisch (Eds.), "Polymer Alloys, Blends, Grafts, and Interpenetrating Networks," Polymer Sci. and Technol., Vol. 10, 1977.
2.  A. Noshay and J. E. McGrath, "Block Copolymers," Academic Press, New York, 1977.
3.  R. J. Ceresa (Ed.), "Block and Graft Copolymerization," Vol. 1, John Wiley and Sons, New York, 1973.

4. J. J. Burke and V. Weiss (Eds.), "Block and Graft Copolymers," Syracuse Press, Inc., Syracuse, New York, 1973.
5. D. C. Allport and W. H. Janes (Eds.), "Block Copolymers," Applied Science Publishers, London, 1973.
6. C. H. Bamford and E. F. T. White, Trans. Fara. Soc., 54:268 (1958).
7. A. Guyot, et al., Inf. Chim., 116:127 (1973).
8. S. N. Gupta and U. Nandi, Macromol. Chem., 176:3179 (1975).
9. D. J. Angier and W. F. Watson, J. Polymer Sci., 25:1 (1957).
10. A. S. Dunn, et al., Trans. Fara. Soc., 50:279 (1954).
11. W. Kawai, J. Polymer Sci., Chem. Ed., 15:1479 (1977).
12. K. F. O'Driscoll and A. U. Sridharan, Appl. Polymer Symp., 26:135 (1975).
13. G. Smets, et al., Makromol. Chem., 23:162 (1953).
14. L. P. H. Chow and I. Piirma, paper presented at American Chemical Society Meeting, 1975; Polymer Preprints, 16(2): 506 (1975).
15. I. Piirma and B. Gunesin, paper presented at American Chemical Society Meeting, 1977; Polymer Preprints, 18(1):687 (1977).
16. J. Hiemeleers and G. Smets, Macromol. Chem., 47:7 (1961).
17. Y. Minoura and Y. Ogata, J. Polymer Sci., A-1, 7:2547 (1969).
18. R. B. Seymour, et al., Advan. Chem. Ser., 99:418 (1971).
19. R. B. Seymour, et al., Appl. Polymer Symp., 26:249 (1975).
20. R. B. Seymour, et al., Appl. Polymer Symp., 25:69 (1974).
21. R. B. Seymour and G. A. Stahl, J. Polymer Sci., Chem. Ed., 14:2545 (1976).
22. R. B. Seymour, et al., Advan. Chem. Ser., 142:309 (1975).
23. L. Gindin, A. Abkin, and S. Medvedev, J. Phys. Chem. USSR, 21:1269 (1974).
24. M. Imoto, et al., Makromol. Chem., 82:277 (1965).
25. T. Ikegami and H. Hirai, J. Polymer Sci., A1, 8:195 (1970).
26. B. K. Patnaik and N. G. Gaylord, J. Macromol. Sci., Chem. A5:1239 (1971).
27. R. B. Seymour, et al., Polymer, 18:1157 (1977).
28. L. M. Minsk and E. W. Taylor, U.S. Pat. 2,582,055 (January 5, 1952).
29. R. B. Seymour and G. A. Stahl, Rev. Sci. Instrum., 46:1467 (1975).
30. N. G. Gaylord and A. Takahashi, Advan Chem., Ser. 1, 91:94 (1969).
31. L. H. Young, "Polymer Handbook," 2nd Edition, Eds. J. Brandrup and E. H. Immergut, Wiley-Interscience, New York, 1975, Chapter II.

COMPARISON OF THE STRUCTURE-PROPERTIES IN 2,4 TDI BASED POLYETHER

POLYURETHANES AND POLYURETHANEUREAS

C.S. Paik Sung

Department of Materials Science and Engineering
Massachusetts Institute of Technology
Cambridge, Massachusetts  02139

INTRODUCTION

The objective of this paper is to review and compare the
structure-property relationships in segmented polyurethanes and
polyurethaneureas based on 2,4 toluene diisocyanate. In this paper,
polyurethanes refer to polymers extended with butane diol, while
polyurethaneureas refer to polymers extended with ethylene diamine.
Both polyurethanes and polyurethaneureas consist of alternating
soft segment (aliphatic polyether) and hard segment (aromatic urethane or urea).  Due to the thermodynamic incompatibility of the
hard segments with the soft segments, microphase segregation occurs
in these polymers to a varying degree, which is strongly influenced
by the compositional variables (1-4).  In this paper, the effect of
urethane linkage (as in polyurethanes) versus urea linkage (as in
polyurethaneureas) in the hard segment will be discussed with polymers based on asymmetric diisocyanates, such as 2,4 toluene diisocyanate.  Due to its asymmetry, the hard segment domains in these
polymers are amorphous (2,5), thus complications arising from a
partial crystallinity in the hard segment domains (as in MDI based
polyurethanes) are absent. The soft segment used in these polymers
is polytetramethylene oxide (M.W. 1000 or M.W. 2000), which was
found to be amorphous in segmented polyurethanes or polyurethaneureas. Therefore, the polymers in this study are amorphous both in
the hard segment domain and in the rubbery soft segment, which makes
it easier to correlate the structure of the polymer (e.g. phase segregation, hydrogen bonding and morphology) with the polymer properties.  Another variable which was studied was the length of the
hard segment (or total urethane or urea content).  In this study,
we will report on the results of the thermal transition properties,

119

small angle x-ray scattering studies, IR studies, and mechanical
properties.

EXPERIMENTAL

## Polymer Syntheses

2,4 Toluene diisocyanate (2,4 TDI) was obtained from Aldrich
Chemical Company and vacuum distilled. Poly(tetramethylene oxide)
(PTMO) of molecular weight 1000 and 2000 were from Quaker Oats
Company. Anhydrous 1,4-butanediol (BD) was from General Aniline
and Film, while anhydrous ethylene diamine was obtained from
Fisher Scientific Company. Polyurethanes were made based on the
two-step bulk polymerization procedure outlined by Pigott and coworkers (6) where prepolymer was first made by end-capping PTMO with
diisocyanate and butanediol was added in the second step. Polyurethaneureas were synthesized according to the method described
by Lyman and coworkers (7), which involved the solution polymerization technique due to the high reactivity of aliphatic diamine chain
extender. First, the solution of polyether was prepared in 1:1 mixture of dimethyl sulfoxide and methyl isobutyl ketone and 2,4 TDI
was added to the solution. The prepolymer solution was stirred at
100°C for 1.5 hours and ethylene diamine was added and stirred for
an additional hour at 30-40°C. The polymer was precipitated and
dried in a vacuum oven at 50-60°C for at least a week.

For both polyurethanes and polyurethaneureas, the molar ratio
of 2,4 TDI, butanediol (or ethylene diamine) and PTMO was varied
in five equal steps from 2:1:1 to 6:5:1. For polyurethanes, 5%
excess diisocyanate was used while for polyurethaneureas, the exact stoichiometry was used.

## Polymer Characterization

Thermal Analysis. For DSC analysis, polyurethane samples were
run either as compression molded at 160°C or as-polymerized bulk,
while polyurethaneurea samples were run as-polymerized bulk. Differential scanning calorimetry was carried out on a Perkin-Elmer
DSC II, equipped with the scanning auto zero accessory, at a heating rate of 20°C/min. The sensitivity chosen was either 2 mcal/
sec or 5 mcal/sec. For low temperature scans ranging from 160°K
to 350°K, helium was used as a carrier gas, while argon was used for
high temperature scans. The sample weight was approximately 10~25mg.

IR Studies. Thin polymer films (~2μ in thickness) were cast
directly on NaCl plates from 1% DMF solution and dried in a vacuum
oven until the IR spectra did not show any evidence of residual
solvent. A Beckman 12 infrared spectrophotometer was used for

polyurethanes, while a Fourier Transform IR Spectrometer (Digilab
FTS-14) was used for polyurethaneureas.  All the spectra were recorded at room temperature.

Small Angle X-Ray Scattering Studies.  For polyurethanes, a
Kratky camera was used while for polyurethaneureas, SAXS patterns
were initially recorded in photographic films and a densitometer
was used to scan the intensity as a function of angle.  In order
to verify that the two techniques provide similar results, one of
the polyurethanes (2,4 TDI-BD-PTMO 1000 (2:1:1)) was studied by
the photographic method.  The densitometer scan of SAXS profile
was exactly the same as observed by a Kratky camera.

Mechanical Properties.  Test specimens were prepared by casting
films from 5% DMF solution in DMF onto clean glass plates and drying them in a vacuum oven for a week.  Tensile tests were carried
out at a strain rate of 1.2%/sec with an Instron Model 1122 using
a 2kg load cell.  Hysteresis measurements were made by loading and
unloading the strip specimens to an increasing strain level at each
cycle.  Strain levels were varied incrementally from 25% to 1050%.
The percent hysteresis for a given cycle is calculated by taking the
ratio of the area bounded by the loading-unloading curves to the
total area under the loading curve.

Molecular Weight Determination.  For polyurethanes, the number
average molecular weight was estimated by vapor phase osomometry
(Perkin-Elmer-Coleman Model 115) on a 2% polymer solution in dimethyl formamide.  For polyurethaneureas, number average molecular
weight was estimated by GPC with a dilute polymer solution (0.25%
in DMF, which contains 0.05M LiBr to prevent aggregation).

RESULTS AND DISCUSSION

Schematic I shows the chemical structure of polyether polyurethanes and polyether polyurethaneureas.  It is noted that in
polyether polyurethanes, urethane (NHCOO) is the repeat unit in
the hard segment domain, while in polyether polyurethaneureas, the
urea (NHCONH) is the repeat unit in the hard segment domain and
the urethane linkage is only found as a bond to connect the hard
segment domain and the soft segment phase, i.e., at the interface.
The major difference between these two series is then the presence
of urea linkage versus urethane linkage in the hard segment domain.
One may be concerned about the effect of butyl group in polyurethanes versus ethyl group in polyurethaneureas.  In the work reported by Bonart and coworkers (8), the hard segment softening
point  in MDI-ED-PBA 2000 (polyester) series was 187°C, while that
for MDI-BD (butylene diamine)-PBA 2000 series was 179°C.  This result implies that in polyurethaneureas, the effect of changing the
chain extender from ethylene diamine to butylene diamine is small.

|← Hard Segment ──────────→|← Interface ──────→|← Soft Segment→|

2,4 TDI-BD-PTMO Series
(Polyether Polyurethanes)

2,4 TDI-ED-PTMO Series
(Polyether Polyurethaneureas)

Schematic I.
Chemical Structure of Polyether Polyurethanes
and Polyether Polyurethaneureas

Therefore, we will assume that the major difference between polyurethanes and polyurethaneureas, in our study, is due to the difference of urea linkage versus urethane linkage.

Table 1 shows the molecular weight and the total urethane or
urea content in each composition for the polymers under study.

Thermal Transition Properties.  Thermal transition properties
of 2,4 TDI-BD-PTMO 1000 were found to be a strong function of urethane content (9), as shown in Figure 1(a) and the data of Table 1.
$T_g$ of the soft segment phase varies from -36°C to 23°C with increasing urethane concentration.  This corresponds to an elevation
from the $T_g$ value of the free soft segment (-85°C), in the range of
49 to 108°C.  The increase in $T_g$ explains the progressive change in
properties from soft and tacky to rubbery and, finally, plastic.

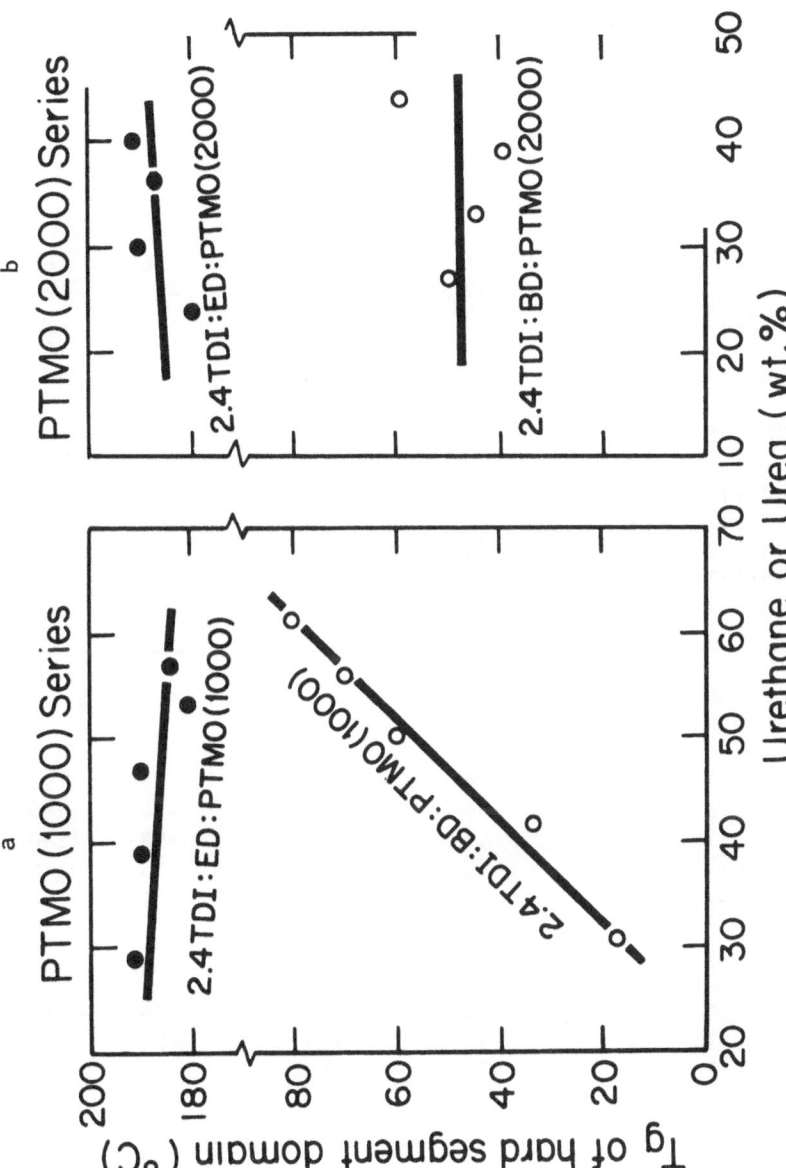

Fig. 2. Comparison of Tg of the hard segment domain in polyether polyurethanes and polyether polyurethaneureas.

Table 1.  Properties and Thermal Transitions of 2,4 TDI
Based Polyether Polyurethanes and Polyether Polyurethaneureas

| Sample | Urethane or Urea Content | M.W. x10$^{-3}$ | $Tg_1$* | $Tg_2$** |
|---|---|---|---|---|
| **Polyurethanes** | | | | |
| **2,4 TDI-BD-PTMO 1000** | | | | |
| 2:1:1 | 31 | 16 | -36 | 18 |
| 3:2:1 | 42 | 14 | -14 | 33 |
| 4:3:1 | 50 | 13 | 1 | 60 |
| 5:4:1 | 56 | 25 | 14 | 70 |
| 6:5:1 | 61 | 14 | 23 | 80 |
| **2,4 TDI-BD-PTMO 2000** | | | | |
| 2:1:1 | 19 | | -67 | |
| 3:2:1 | 27 | | -65 | 50 |
| 4:3:1 | 33 | | -69 | 45 |
| 5:4:1 | 39 | | -72 | 40 |
| 6:5:1 | 26 | | -70 | 60 |
| **Polyurethaneureas** | | | | |
| **2,4 TDI-ED-PTMO 1000** | | | | |
| 2:1:1 | 29 | 21 | -53 | 192 |
| 3:2:1 | 39 | 28 | -54 | 190 |
| 4:3:1 | 47 | 33 | -58 | 190 |
| 5:4:1 | 53 | 43 | -55 | 180 |
| 6:5:1 | 57 | 27 | -61 | 184 |
| **2,4 TDI-ED-PTMO 2000** | | | | |
| 3:2:1 | 24 | 36 | -76 | 180 |
| 4:3:1 | 30 | 34 | -74 | 190 |
| 5:4:1 | 36 | 35 | -73 | 186 |
| 6:5:1 | 40 | 25 | -78 | 191 |

\*    Corresponds to the $T_g$ of the soft segment phase.

\*\*  Corresponds to the $T_g$ of the hard segment domain.

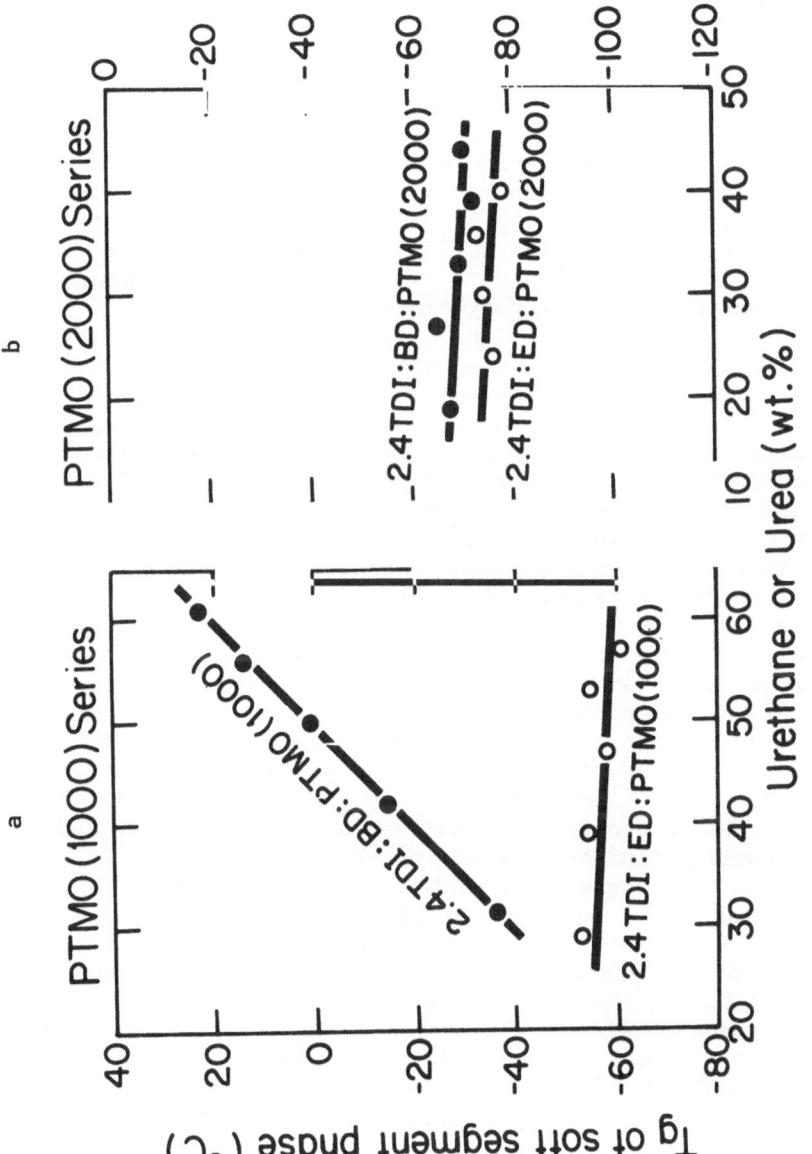

Fig. 1.  Comparison of T$_g$ of the soft segment phase in polyether polyurethanes and polyether polyurethaneureas.

Similar behavior was shown for an intermediate temperature transition observed by thermomechanical analysis in the range of 18°C to 80°C, as shown in Figure 2(a) and the data of Table 1. This transition is interpreted as the glass transition temperature of the amorphous hard segment domains whose structure improves with increasing hard segment length, which is proportional to the urethane content. A higher temperature transition was detected only in the samples of highest urethane content, and then only on the initial heating. This transition is believed to result from some allophonate or biuret bonding, which arises from the small excess of diisocyanate in the polymerization recipe. Therefore, in 2,4 TDI-BD-PTMO 1000 polyurethane series, the phase segregation is rather poor and the extent of phase mixing is extensive. In comparison, the thermal properties of 2,4 TDI-ED-PTMO 1000 polyurethaneurea series were independent of urea content. The $T_g$ of the soft segment phase was much lower than that of polyurethanes, i.e., around -55°C as shown in Figure 1(a) and the data of Table 1. This $T_g$ is elevated by only 30°C from the $T_g$ of the free soft segment (-85°C). Even though this result means the presence of some solubilized hard segment in the soft segment phase, the amount of mixing is expected to be much smaller than in the polyurethanes. In addition to the low $T_g$, around -55°C, all the samples in 2,4 TDI-ED-PTMO 1000 polyurethaneureas exhibit a very high $T_g$ of the hard segment domain, around 190°C as shown in Figure 2(a) and the data of Table 1. This behavior indicates that the phase segregation in polymers extended with ethylene diamine is greatly improved.

When the length of the soft segment is increased from M.W. 1000 to M.W. 2000, improvement in phase segregation has been observed in both polyurethanes and polyurethaneureas. As shown in Figure 1(b) and the data of Table 1, the $T_g$ of the soft segment of 2,4 TDI-BD-PTMO 2000 polymers are quite low (around -60°C) and found to be independent of urethane content (10). In these polymers, the soft segment remains as amorphous as evidenced by DSC and wide angle x-ray diffraction studies in spite of the high molecular weight. The $T_g$ of the hard segment domain was observed around 50°C. This means that the hard segment domain may contain a substantial amount of the solubilized soft segment since the $T_g$ of pure hard segment is expected to be around 100°C (11). Therefore, in 2,4 TDI-BD-PTMO 2000 polyurethanes, the soft segment phase is relatively pure, but the hard segment domain is mixed. As a result, these polymers provide good low temperature flexibility, but rather poor thermal stability. In contrast, 2,4 TDI-ED-PTMO 2000 polyurethaneureas provide excellent low temperature flexibility and thermal stability, since the $T_g$ of the soft segment is very close (-75°C) to the $T_g$ of the free soft segment, as shown in Figure 1(b), and the $T_g$ of the hard segment domain is much higher (180°C) than that of the analogous polyurethanes, as shown in Figure 2(b) (2). Since the $T_g$ of the soft segment phase in 2,4 TDI-ED-PTMO 2000 series is only elevated by 10°C, which can be attributed to the tying-in of the end groups

of the polyether soft segment, we can assume that the amount of
solubilized hard segment in the soft segment phase is negligible.

Small Angle X-Ray Scattering Studies. In order to provide independent evidence of domain structure and to compare domain organization between polyurethanes and polyurethaneureas, SAXS patterns
were obtained. Figure 3(a) shows the angular dependence of intensity
obtained at room temperature for 2,4 TDI-BD-PTMO 1000 polyurethanes
(12). It is clear that the three polymers with longer hard segment
exhibit the type of angular dependence which implies the presence of
domain structure. It is also to be noted that total area under the
curve is strongly dependent on the urethane content. Since the integrated scattering intensity is proportional to the product of
three terms representing the electron density difference and the
weight fractions of the dispersed and the continuous phase, the results imply that the extent of phase segregation increases with increasing urethane content, perhaps accompanied by some improving
domain organization. On the other hand, quite a different trend is
observed in SAXS patterns of 2,4 TDI-ED-PTMO 1000 polyurethaneureas,
as shown in Figure 3(b) (2). All the samples in this series, even
the low urea content composition, such as a 2:1:1 composition, exhibit the type of angular dependence which implies the presence of
domain structure. Furthermore, the area under each scattering curve
is not much different, even though polymers with higher urea content
tend to show somewhat greater area under the curve. This result
seems to suggest that the organization of domains are comparable
within this series. This result is consistent with the thermal
transition properties. It is expected that the SAXS patterns for
both 2,4 TDI-BD-PTMO 2000 polyurethanes and 2,4 TDI-ED-PTMO 2000
polyurethaneureas will show strong evidence of well-organized domains. A quantitative analysis concerning domain organization and
interfaces based on SAXS results are now in progress in collabora-
ation with Professor G. Wilkes.

IR Studies. Since polyurethanes and polyurethaneureas are found
to be extensively hydrogen bonded, an analysis of IR spectra provides an alternate approach of obtaining a more detailed understanding of the relation between composition, transition behavior
and properties. The infrared analysis depends on the resolution
of NH and carbonyl band into the bonded and non-bonded components.
In polyurethanes, since the proton acceptors are mainly the carbonyl group of the urethane hard segment and the oxygen of the
polyether soft segment, the fraction of bonded carbonyl can be
taken as a measure of the extent of phase segregation. In turn,
the fraction of NH groups bonded to ether, determined by difference,
indicates the degree of hard segment-soft segment mixing. Figure
4(a) illustrates the IR spectra in the regions of NH stretching
and carbonyl stretching for 2,4 TDI-BD-PTMO 1000 polyurethanes (11).
It is well known that a band near 3300 cm$^{-1}$ is caused by the H-
bonded NH groups, while the free NH group appears near 3460 cm$^{-1}$.

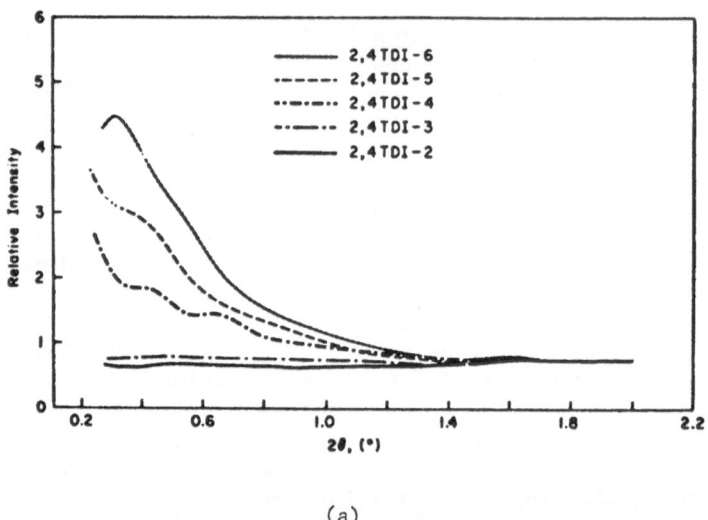

(a)

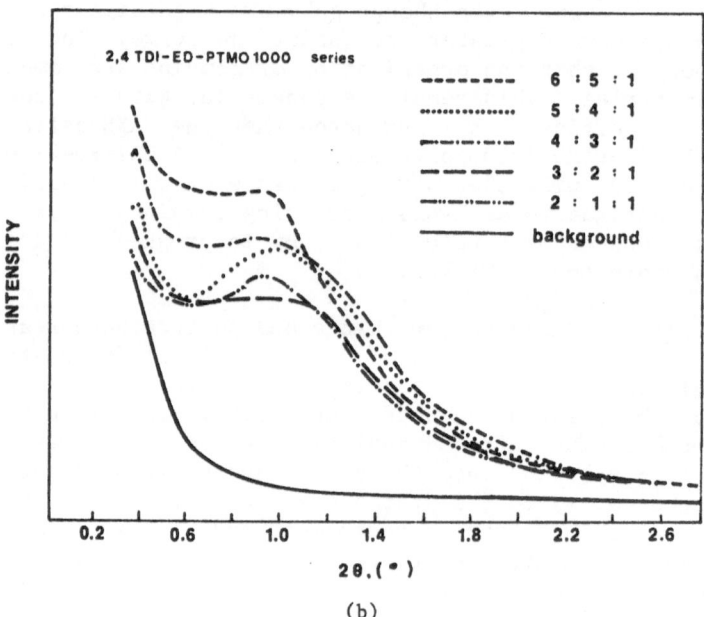

(b)

Fig. 3.   Angular dependence of SAXS intensity (a) 2,4 TDI-BD-PTMO
1000 series, (b) 2,4 TDI-ED-PTMO 1000 series.

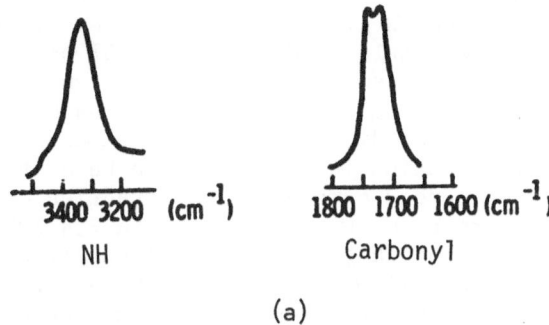

NH                    Carbonyl

(a)

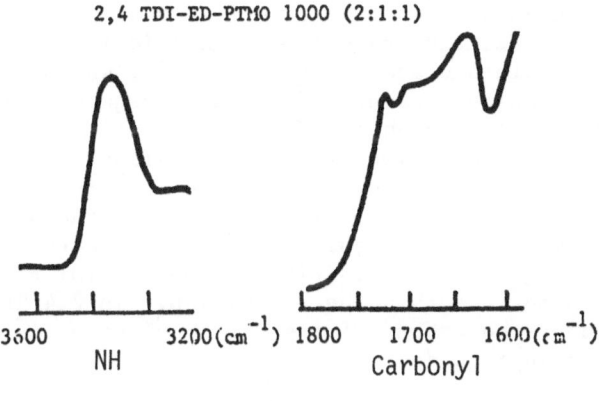

2,4 TDI-ED-PTMO 1000 (2:1:1)

NH                    Carbonyl

(b)

Fig. 4.  IR spectra in the region of NH and carbonyl for (a) 2,4
         TDI-BD-PTMO 1000 (2:1:1) polymer, (b) 2,4 TDI-ED-PTMO 1000
         (2:1:1) polymer.

MacKnight and Yang have reported an integrated extinction coef-
ficient $E_f = 3.44 \times 10^3$ $\ell mol^{-1}$ $cm^{-1}$ for the free NH group and $E_b =$
$1.19 \times 10^4$ $\ell mol^{-1}$ $cm^{-1}$ for the H-bonded NH groups from measurements
of a model compound (13). Based on these extinction coefficients,
the fraction of H-bonded NH groups was calculated by the curve re-
solving technique. The results indicate that 95% of all NH groups
are hydrogen bonded in the solid state at room temperature. The carbonyl region shows splitting of the absorption band into two peaks
at 1740 and 1720 $cm^{-1}$. Studies on model compounds of n-butyldiure-
thane of 2,4 TDI in solution indicated that a band at 1720 $cm^{-1}$ is
a result of hydrogen bonded carbonyl. The fraction of bonded carbonyl is calculated to be between 46 to 60% in 2,4 TDI-BD-PTMO 1000
series. The remainder of NH groups are necessarily bonded with
other proton acceptors. If we assume that the ether oxygen is the
major proton acceptor, then the rest of the NH groups are bonded
to ether oxygen of the soft segment. This type of hydrogen bonding
between the solubilized hard segment and the soft segment phase is
found to be the cause of elevating $T_g$ of the rubbery phase, beyond
what the copolymer equation would predict. In fact, the hydrogen
bonding of this type acts as if they are cross-links in regards to
the elevation of $T_g$. Additional insight concerning the effect of
hydrogen bonding on the higher thermal transitions has been obtained by analyzing the temperature dependence of the infrared
spectra in the NH and carbonyl regions (12). In 2,4 TDI-BD-PTMO
1000 series, the onset temperature for dissociation of hydrogen
bonded NH is variable and occurs at 40 to 60°C, which is close to
the glass transition temperature of the hard segment domain. Surprisingly, there is little change in the carbonyl region up to 150°C,
implying that the dissociation of NH is due almost entirely to the
disruption of urethane-to-ether bonding. A surprising observation
is the relative stability of the inter-urethane bonding in these
2,4 TDI based samples. This could reflect the greater stability
of hydrogen bonding at the 4 position in the 2,4 TDI ring due to
smaller steric hindrance than the 2 position. In 2,4 TDI-BD-PTMO
2000 polyurethanes the amount of free carbonyl group is less due to
the better phase segregation.

On the other hand, IR spectra of polyurethaneureas (both 2,4
TDI-ED-PTMO 1000 and 2,4 TDI-ED-PTMO 2000 series) reveal more interesting features. Figure 4(b) shows IR spectra of NH regions
and carbonyl regions of 2,4 TDI-ED-PTMO 1000 polyurethaneureas (14).
The IR spectra of 2,4 TDI-ED-PTMO 2000 series is very similar to
that of 2,4 TDI-ED-PTMO 1000 series. In both of these series, the
NH band in the region of 3200-3500 $cm^{-1}$ appears completely hydrogen
bonded, since only a peak at 3300 $cm^{-1}$ is visible, while the free
NH peak at 3460 $cm^{-1}$ is negligible. This was found to be true for
other compositions of both series. In the carbonyl region, more
than half of the urethane carbonyl appear as free carbonyl peak
at 1740 $cm^{-1}$, while the urea carbonyl peak appears as hydrogen

bonded form at 1640 ~1660 cm$^{-1}$. There is negligible amount of free urea carbonyl, which would appear at 1690 cm$^{-1}$. With in- creasing urea content, the amount of bonded urethane carbonyl de- creases, or in other words, the free urethane carbonyl increases.

We can derive at least two important observations from the features of IR spectra of polyurethaneureas. First, the fact that much of urethane carbonyl, which is only present at the inter- face between hard segment domain and soft segment domain is free from hydrogen bonding strongly suggests that the interface must be fairly sharp. Second, both NH and urea carbonyl groups are almost completely hydrogen bonded. Since we can assume that the polymer is, for the most part a linear molecule, there are approximately twice as many NH groups per each urea carbonyl in all the composi- tions. For example, at a 2:1:1 composition, there are 6 molar NH groups, 2 molar urethane carbonyl, and 2 molar urea carbonyl. Since only half of the urethane carbonyl is bonded to presumably 1 molar NH group, the rest of the NH groups (5 molar) should be bonded to 2 molar urea carbonyl and to polyether oxygen. Thermal transition studies reported in the earlier section, indicated that the mixing of hard segment in the soft segment phase is likely to be small in 2,4 TDI-ED-PTMO 1000 series and negligible in 2,4 TDI-ED-PTMO 2000 series. Therefore, we can assume that the hydrogen bonding of NH groups to polyether is small, which means that most of the 5 molar NH groups should be bonded to 2 molar urea carbonyl. The only way that this can be accomplished is for each urea carbonyl to be bonded to two NH groups. This type of 3-D hydrogen bonding should also oc- cur in all compositions in our polyurethaneureas, but to a greater extent in compositions such as 5:4:1 and 6:5:1, since they tend to have longer hard segment lengths.

Mechanical Properties. Even though we have not carried out any quantitative measurements of the mechanical properties in polyue- thanes, it is obvious from the $T_g$ behavior of the soft segment phase in 2,4 TDI-BD-PTMO 1000 series that the polymer mechanical properties change progressively with increasing urethane content. At a 2:1:1 composition, it is a clear and soft, sticky rubber. At a 3:2:1 com- position it is a tough and snappy rubber. But, at a higher urethane content, such as 4:3:1 and 5:4:1, they are elastic but not snappy, and at 6:5:1 composition it is almost plastic. This type of pro- gressive change to an almost plastic behavior is due to the elevated $T_g$ of the soft segment phase in these polymers. When the polyether molecular weight is increased from 1000 to 2000, as in 2,4 TDI-BD- PTMO 2000, the mechanical properties are better than PTMO 1000 series, i.e., PTMO 2000 series are clear, rubbery, and moderately tough polymers.

In polyurethaneureas, the consequences of better phase segre- gation and domain organization in comparison to similar polyure-

thanes are well reflected in the mechanical properties (15). Figure
5(a) shows the stress-strain curves obtained with two samples from
the 2,4 TDI-ED-PTMO 1000 series and one sample from the 2,4 TDI-
ED-PTMO 2000 series.  The tensile modulus for two samples in the
PTMO 1000 series is much greater than that for a PTMO 2000 sample.
This behavior is more clearly shown in Figure 5(b), which represents
the initial portion of the curve with better accuracy.  Tensile
modulus, tensile strength, and elongation at break depend on the
composition of the polymer.  Within the PTMO 1000 series, the tensile modulus and tensile strength were relatively high and they increase with increasing urea content,while elongation at break decreases slightly.  Specifically, with a 2:1:1 composition, the tensile modulus and the tensile strength were 2000 kg/cm$^2$ and 400 kg/
cm$^2$, respectively, and the elongation at break was 500%.  With a
5:4:1 composition, the tensile modulus was 4000 kg/cm$^2$, the tensile
strength was 500 kg/cm$^2$ and the elongation at break was 400%.  With
a 4:3:1 composition of the PTMO 2000 series, which has the same
overall urea content as a 2:1:1 composition of the PTMO 1000 series,
the tensile modulus was about 200 kg/cm$^2$ (one tenth of that of the
latter) and the tensile strength was 250 kg/cm$^2$, which is lower
than that for the PTMO 1000 analog.  However, the elongation at
break is 1100%.  Whereas the two curves corresponding to the 2,4
TDI-ED-PTMO 1000 series show a marked deviation from the behavior
of unfilled vulcanized rubber, the curve corresponding to a 2,4
TDI-ED-PTMO 2000 sample exhibits less degree of the deviation. This
result suggests that the domains in the PTMO 2000 series of polyurethanes are probably spherical and better separated than those in the
PTMO 1000 series.  Also, the PTMO 2000 polymer is found to have
lower hysteresis value as compared with the PTMO 1000 polymer due
to the better separation of domains.  This is shown in Figure 6,
where the percent hysteresis is plotted as a function of strain imposed in an incremental manner.  For both samples of 2,4 TDI-ED-
PTMO 1000 series, the percent hysteresis quickly reaches about 70%
and stays constant between 75-80% even with a futher increase in
the elongation.  There is no difference in the curve between a
2:1:1 composition and a 5:4:1 composition, which contains about
twice the total urea content.  In contrast, the hysteresis value
in the 2,4 TDI-ED-PTMO 2000 series is distinctly lower than that
of two members of the PTMO 1000 series.  A 4:3:1 composition of
PTMO 2000 series, which has the same overall urea content as a
2:1:1 composition of PTMO 1000 series, exhibits a much lower hysteresis.  As illustrated in Figure 6, the initial hysteresis value
for this sample of PTMO 2000 series is around 30% and increases
slowly with a further elongation up to 65-70% at a 1050% elongation.


CONCLUSION

    We have compared the effect of urethane linkage versus that of
urea linkage in the hard segment on the structure and properties

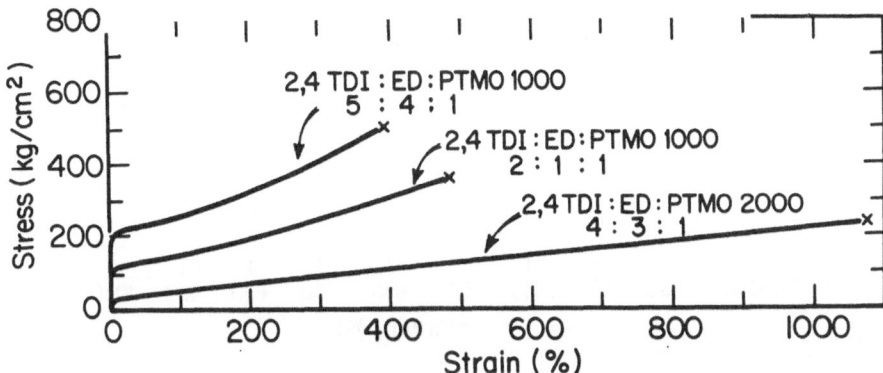

Fig. 5.(a).   Stress-strain behavior of polyether polyurethaneurea
              elastomers (strain rate = 1.2%/sec.).

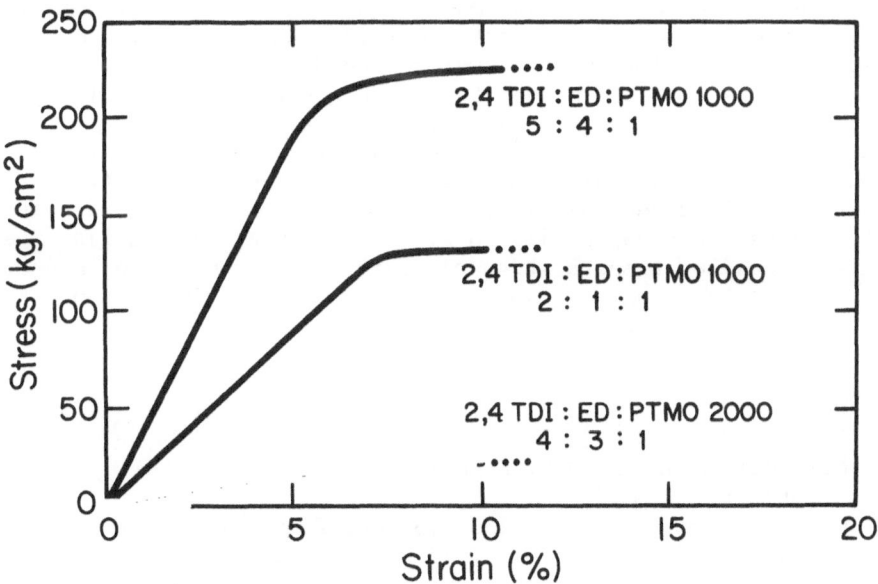

Fig. 5(b).   Initial portion of the stress-strain behavior of poly
             ether polyurethaneurea elastomers (strain rate =
             1.2%/sec.).

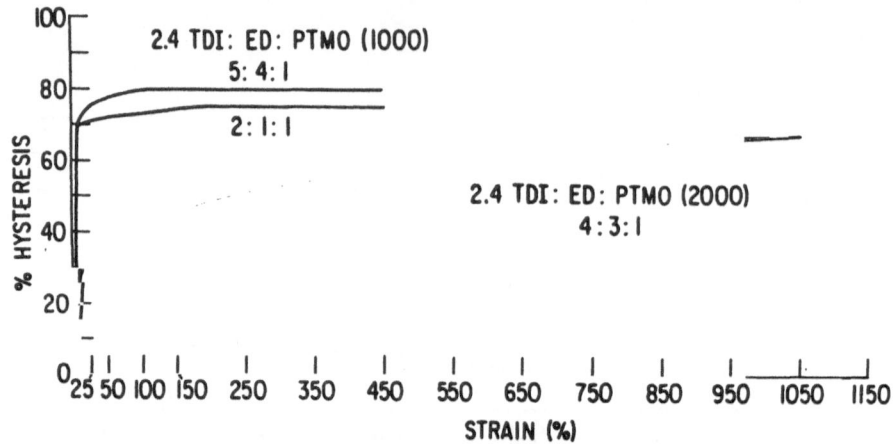

Fig. 6.    Percent hysteresis as a function of strain for
           polyurethaneureas.

of 2,4 TDI based polyether polyurethanes and polyurethaneureas.
The effect of urea linkage is to drastically improve phase segregation so that the polyurethaneureas demonstrate lower $T_g$ of the
rubbery soft segment phase and much higher $T_g$ of the hard segment
domain, even in low urea content.  Our IR studies indicate the
presence of three dimensional hydrogen bond in which one urea carbonyl group is hydrogen bonded to two NH groups in the hard segment
domain of polyurethaneureas, even in low urea content.  This type
of hydrogen bonding might be responsible for the drastic improvement
in phase segregation. Both phase segregation and domain structure
lead to excellent mechanical properties, as demonstrated in polyurethaneureas, and especially in 2,4 TDI-ED-PTMO 2000 series, i.e.,
moderate tensile strength, high elongation at break, lower hysteresis, and greater toughness value (as expressed by the area under
the stress-strain curve).

REFERENCES

1.    C.S. Paik Sung and N.S. Schneider, J. Materials Sci., 13, 1689
         (1978).
2.    C.S. Paik Sung, C.B. Hu and C.S. Wu, Polym. Prepr. Am. Chem.
         Soc., Div. Polym. Chem., 19-2, 679 (1978).
3.    N.H. Ng, A.E. Allegrezza, R.W. Seymour and S.C. Cooper, Polymer,
         14, 255 (1973).
4.    Y.P. Chang and G.L. Wilkes, J. Polym. Sci. Polym. Physics, 13,
         455 (1975).
5.    C.S. Paik Sung, C.S. Wu and C.B. Hu, Polym. Prepr. Am. Chem.

Soc., Div. Polym. Chem., 19-2, 686 (1978).

6.   K.A. Pigott, B.F. Frye, K.R. Allen, S. Steingiser, W.C. Dan, J.H. Saunders and E.E. Hardy, J. Chem. Eng. Data, 5, 391 (1960).

7.   D.J. Lyman, D.W. Hill, R.K. Stirk, C. Adamson and B.R. Mooney, Trans. Amer. Soc. Artif. Int. Organs, 18, 19 (1972).

8.   R. Bonart, L. Morbitzer and H. Rinke, Kolloid-Z. U.Z. Polymere, 240, 807 (1970).

9.   N.S. Schneider, C.S. Paik Sung, R.W. Matton and J.L. Illinger, Macromolecules, 8, 62 (1975).

10.  N.S. Schneider and C.S. Paik Sung, Polym. Eng. & Sci., 17-2, 73 (1977).

11.  C.S. Paik Sung and N.S. Schneider, Macromolecules, 8, 68 (1975).

12.  C.S. Paik Sung and N.S. Schneider, Macromolecules, 10, 452 (1977).

13.  W.J. MacKnight and M. Yang, J. Polym. Sci., Part C, No. 42, 817 (1973).

14.  C.S. Paik Sung, C.B. Hu and T.W. Smith, Polym. Prepr. Am. Chem. Soc., Div. Polym. Chem., 19-2, 692 (1978).

15.  C.S. Paik Sung, T.W. Smith, C.B. Hu and N.H. Sung, Macromolecules, 12, 538 (1979).

ACKNOWLEDGEMENT

The generous support of the Whitaker Health Sciences Fund is greatly appreciated. The author also thanks Dr. N.S. Schneider of Army Materials and Mechanics Research Center and Mr. C.B. Hu and Mr. T. W. Smith of M.I.T.

THE PREPARATION AND PHOTOOXIDATIVE DEGRADATION OF POLYESTERURETHANE-

POLY(METHYL METHACRYLATE) TRIBLOCK COPOLYMERS

John A. Simms

E. I. du Pont de Nemours & Company, Inc.
E-174/308, Wilmington, De. 19898

INTRODUCTION

Block copolymers of the ABA type made from hydrocarbon monomers by anionic polymerization have been extensively investigated (1, 2). Much less has been done with nonhydrocarbon block polymers. Such systems as have been reported show even broader temperature ranges of technical utility than the more familiar styrene/diene/styrene systems (3, 4, 5). It is the purpose of this paper to describe a process for preparing polymethyl methacrylate containing triblock copolymers. The physical property balances that can be gotten and the photooxidative degradation processes that occur will also be covered.

This work was underway at the time J. H. Grezlak and G. L. Wilkes (3) published on a similar synthetic approach to block copolymers. In their process a preformed, isocyanate terminated polyester-urethane is reacted with monohydroxyfunctional polymethyl methacrylate. In the new procedure reported here a monohydroxyfunctional polymethyl methacrylate-polyester diol mixture is azeotropically dried in toluene solution and then simultaneously extended and end capped by reaction with diisocyanate. This approach has the advantage of simplicity and avoids unnecessary manipulations of moisture sensitive materials.

DISCUSSION

Polymer Synthesis

Four block copolymers were prepared. Polymer I is described in Fig. 1. Polymer II used 3000 $M_N$ neopentyl glycol adipate in the B segment and the same diisocyanate [bis(4,4'-isocyanatocyclo-

137

hexyl) methane] as Polymer I.  This modification was shown to be
much more resistant to hydrolysis than Polymer I.   In addition, UV
caused cyclic decomposition of the ester would be prevented by the
neopentyl structure.   Polymers III and IV used the same polyester
as I but were extended with 2,4-toluenediisocyanate and  $\alpha,\alpha,\alpha',\alpha'$-
tetramethylxylene diisocyanate respectively.   The changes in III
and IV were expected to prevent photooxidative attack at the carbon
next to the -NH- of the urethane linkage.   Initial elongations were
high in all these polymers.

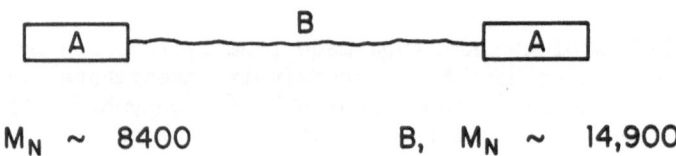

A,  $M_N$ ~ 8400                         B,  $M_N$ ~ 14,900

A  *is*  $H \left( CH_2-\underset{\underset{CO_2CH_3}{|}}{\overset{\overset{CH_3}{|}}{C}} \right)_{84} SCH_2CH_2OH$

B  *contains*   4   3400 $M_N$  ETHYLENEGLYCOL / 1, 4-BUTANEDIOL
ADIPATE SEDMENTS AND

5 OCN—CH$\underset{CH_2-CH_2}{\overset{CH_2-CH_2}{<}}$C—CH$_2$—CH$\underset{CH_2-CH_2}{\overset{CH_2-CH_2}{<}}$CH—NCO

Fig. 1.   Description of a Typical Acrylic/Polyesterurethane
         ABA Block Copolymer[a]

a.   Ten (10) urethane links are present in the  average molecule.
     The polyester contains 58 mole % ethylene glycol and 42 mole
     % 1,4-butanediol.

The polymethyl methacrylate hard block ("A") preparation will be considered first, then the extension and capping process will be described. Attention will be focused in turn on end group structure of the "A" block, stoiometry of "A" block preparation and urethane formation during the extension process.

In the preparation of monohydroxypolymethyl methacrylate described in Fig. 2, 2-mercaptoethanol provides the terminal hydroxyl group. Azobisisobutyronitrile (AIBN) derived initiator fragments make polymer without hydroxyl functionality, but this apparently does not interfere with the development of a high level of physical properties in the completed ABA block polymer.

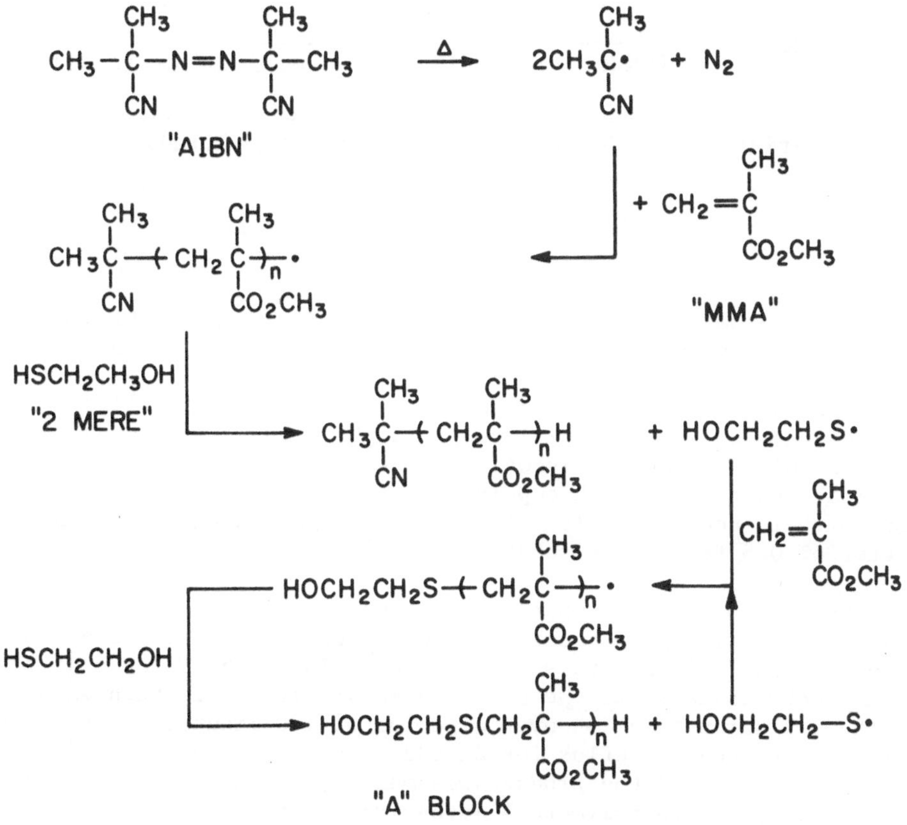

Fig. 2.   Chain Transfer Process for Preparing
          Monohydroxy Poly(methyl methacrylate)
          "A" Blocks

Table 1.   Stoiometry in "A" Block Synthesis

| MMA, moles | AIBN, moles | 2-MERE, moles | Calcd.* $M_N$ | Obs $M_N$ | Ratio Calc/Obs |
|---|---|---|---|---|---|
| 530 | 0.5 | 5.0 | 11,200 | 12,600 | 0.89 |
| 480 | 0.5 | 5.3 | 9,700 | 10,100 | 0.96 |
| 390 | 0.5 | 5.6 | 6,700 | 7,400 | 0.91 |

$$^*M_N = \frac{\text{Wt. MMA}}{2(\text{moles AIBN})(0.5) + (\text{moles 2-MERE})(0.85)}$$

0.5 and 0.85 Are Based On Considerations of Efficiency (AIBN) of Initiation and Completeness of Reaction (2-Mere).

The stoiometry described in Table 1 supports the estimate that about 15% of the polymethyl methacrylate is not hydroxyl modified.

The equation given in Table 1 is quite useful in designing "A" segments of different molecular weight.  This is the key to the control of solution viscosity and spray solids in finishes applications of the block copolymers.  Radical combination to form tetramethyl succinonitrile is the likely cause of the efficiency factor of 0.5 for azobisisobutyronitrile.

The "A" block is prepared by mixing the mercaptoethanol, two thirds of the methyl methacrylate and enough toluene to give a 68% solution, and bringing this mixture to reflux.  The balance of the methyl methacrylate and the azobisisobutyronitrile is then added to the refluxing solution over a period of 160 minutes and the re-action continued at reflux for an additional 120 minutes.  The molecular weight of the product was measured using size exclusion chromatography with polymethyl methacrylate standards.  Conversion was at least 99% in all cases.

The block copolymer synthesis uses monohydroxy terminated polymethyl methacrylate to stop the chain extension of a mixture of polyesterdiol and diisocyanate.  This products a PMMA/polyester urethane/PMMA block polymer.

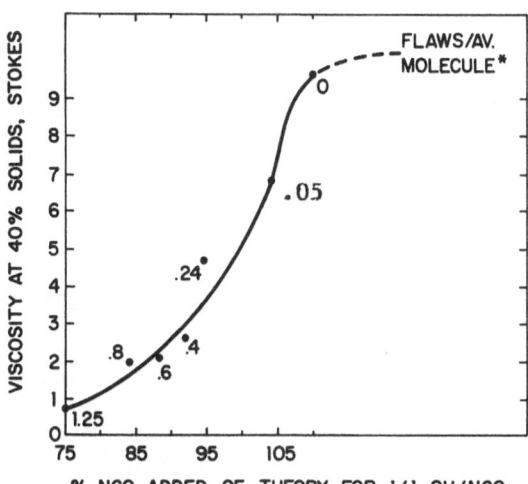

*AVERAGE MOLECULE CONTAINS IO URETHANE LINKS

Fig. 3.   Change in Viscosity as Polymer I is Prepared by Adding
          Bis(4,4'-isocyanatocyclohexyl) Methane to a Mixture
          of Monohydroxypoly(methyl methacrylate) and Poly-
          esterdiol

Fig. 3 shows the build up in viscosity that occurs as Polymer I
is prepared.  As the stoiometric amount of diisocyanate is reached
the viscosity change ends and the characteristic isocyanate bond
at 2250 cm⁻¹ is observed in the infrared spectrum.

The extension process illustrated in Fig. 3 occurs readily at
110°C in the presence of 0.04% stannous octoate (based on non-
volatile content of the charge).  The completion of the extension
process can be considered to be the equivalent of the elimination
of flaws in the molecule.  The numbers along the curve in Fig. 3
refer to moles of diisocyanate which must be reacted to complete
a mole of ABA block polymer.

Completion of the extension process is necessary if the product
is to develop high elongation, Table 2.  The concept that stoiometry
determines the number of flaws in the average molecule  is also
illustrated in Table 2.

Reaction mixtures sampled at less than 75% of stoiometry de-
posited weak, waxy, opaque films from solution.  If the end capping
"A" block was omitted from the solution low modulus elastomers with

Table 2.   Development of Tensile Elongation to Break Polymer I
during the Extension Process[a]

| % NCO ADDED | "FLAWS"/AV MOLECULE | INH. VISC. | % ELONGATION TO BREAK, 23°C |
|---|---|---|---|
| 75 | 1.25 | 0.26 | 4 |
| 88 | 0.6 | 0.31 | 60 |
| 95 | 0.24 | 0.39 | 340 |
| 103 | 0.0 | 0.42 | 620 |

a.   Inherent viscosity was determined at 0.5% concentration
(w/w) 1,2-dichloroethane at 30°C.

inherent viscosities > 2 could be obtained.

The same dependence of tensile elongation characteristics on
block copolymer structure that was observed in this work has been
reported for styrene/butadiene block copolymers, Table 3.   The
importance of the hard/soft/hard sequence is evident since the hard/
soft (A/B) sequence has very low elongation.   This corresponds in
this work to incompletely extended polymer, Table 2.

Structure/Property Correlations
The soft segment content is a major variable in this system,
Table 4, as it is in the case of polystyrene/polybutadiene/poly-
styrene triblock polymer.   A comparison of the data in Tables 3
and 4 shows that at the same "B" segment content these polymers
have similar tensile elongations at break, even though their
chemical composition is in no way similar.   The arrangement of
glassy and rubbery segments and their concentrations dominate.

Transmission electron microscopy showed Polymer I, as cast
from toluene and dried at 121°C for 30 minutes, to have the two
phase structure shown in Fig. 4.   The diagonal lines extending
across the photograph are sectioning artifacts.   The polymer shows
two continuous phases.

Table 3.  Block Copolymer[a]/Property Correlations from
          the Literature[b]

| % BUTADIENE | STRUCTURE | % ELONGATION AT BREAK |
|:---:|:---:|:---:|
| 40 | A / B | 2 |
|  | A / B / A | 70 |
| 47 | A / B / A | 550 |

Ref.  G. Holden, E.T. Bishop and N.R. Legge  — 1969

E. Fischer    1968

M. Matsuo    1968

a.  A = Polystyrene Block, B = Polybutadiene Block
b.  Reference to the original paper and more discussion
    can be found in (1).

Table 4.  Tensile Properties of Polyesterurethane–Poly(methyl
          Methacrylate)[a] Triblock Polymers as a Function of
          Soft Segment ("B") Content*

| % B* | T°C | % ELONGATION AT BREAK | $Kg/cm^2 \times 10^{-2}$ TENSILE STRENGTH | INITIAL MODULUS |
|:---:|:---:|:---:|:---:|:---:|
| 60 | 23 | 810 | 1.6 | 4 |
|  | −30 | 330 | 4.4 | 27 |
| 47** | 23 | 620 | 2.3 | 23 |
|  | −30 | 40 | 4.7 | 79 |
| 40 | 23 | 160 | 2.2 | 72 |
|  | −30 | 5 | 4.0 | 134 |

*3400 $M_N$   ETHYLENEGLYCOL / 1, 4–BUTANEDIOL
            ADIPATE  EXTENDED  WITH  BIS (4, 4'–
            ISOCYANATOCYCLOHEXYL) METHANE

** POLYMER I

a.  PMMA "A" Blocks are 8400 $M_n$

POLYMER - I

$B = 3400\ M_N$ ETHYLENEGLYCOL /1,4-BUTANEDIOL
ADIPATE EXTENDED WITH
BIS (4,4'-ISOCYANATOCYCLOHEXYL) METHANE

$\longmapsto = 1\mu$

Fig. 4.  Transmission Electron Micrograph of
A/B/A Block Copolymer I.

Polymer II, which uses a different polyester diol from
Polymer I, but is very similar in other respects has an included
second phase which appears in the transmission electron micrograph, Fig. 5, to be about 0.25 μ in diameter.  The near vertical
folds extending across the photograph are sectioning artifacts.

Films from Polymer I and II are transparent.  Phase growth
does not occur upon heating as is frequently seen with polymer
mixtures because of the covalent bonding between A and B segments.

Torsion modulus experiments, Fig. 6, showed that Polymer I
has glass transition temperatures at about −45°C and +75°C.  This
is consistent with its two phase, block copolymer structure.

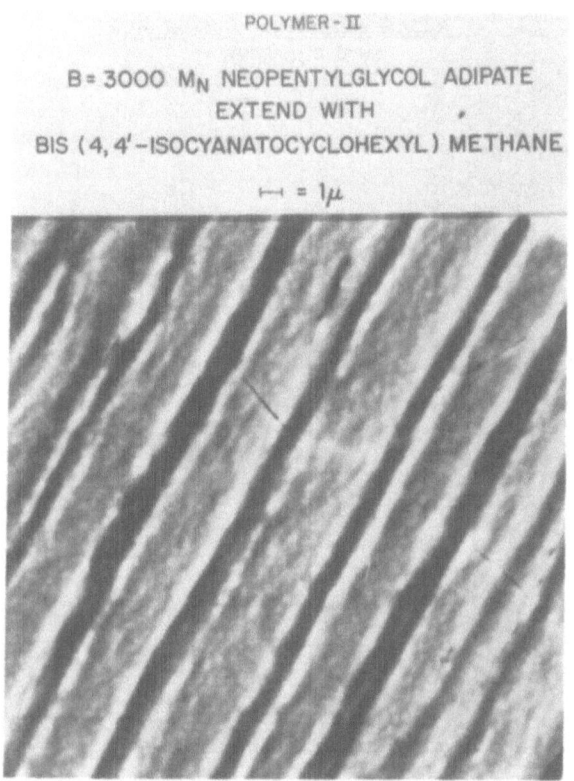

Fig. 5.  Transmission Electron Micrograph of
A/B/A Block Copolymer II

        The 75°C transition in Fig. 6 is due to the glassy polymethyl
methacrylate becoming rubbery.  This rather low transition (PMMA
is usually reported to have a Tg of 105°C) is due to the molecular
weight of this segment (6).  The -45°C transition is consistent
with the soft segment structure.  These results indicate that there
is very little phase mixing in this block copolymer.

Degradation Mechanism
        Loss of tensile elongation, Table 5, in simulated weathering
of 50 micron films is rapid.  The films degrade in 50 hours to the
elongation which they should not reach for 500 hours if they are
to be useful as exterior finishes.  In this experiment the light
source is filtered to remove UV light below 290 nm in wavelength.
Intermittent water spray can be used.  The surface of the 50 micron
thick film is about 60°C during accelerated weathering.

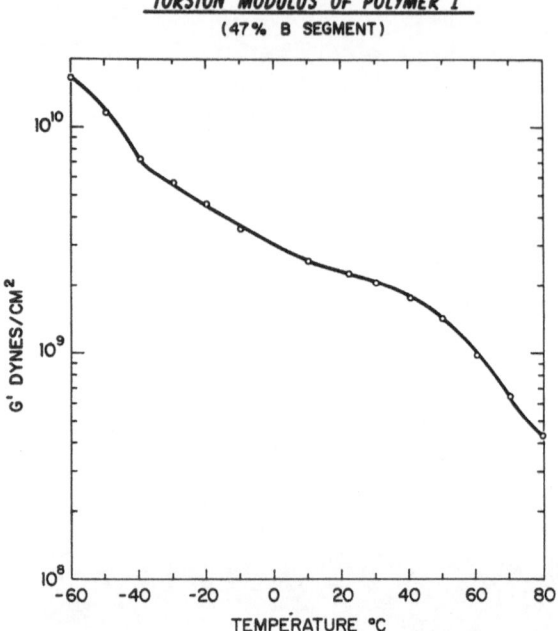

Fig. 6.   Change in Stiffness of an A/B/A Block
          Copolymer[a] with Temperature

a.   The "A" segment is poly(methyl methacrylate)
     of $M_n$ = 8400.  The "B" segment is the polyester-
     urethane from 3400 $M_n$ ethylene glycol/1,4-
     butanediol adipate and bis(4,4'-isocyanato-
     cyclohexyl) methane

     Polymer I was found to lose all of its elongation after three
winter months of 5° South exposure in Florida.

     Oven aging of 50 micron films of Polymer I in air at 82°C
caused no loss of elongation after 1 month, demonstrating the importance of light to the process.  Although water accelerates degradation, it is not necessary.  Thus the major degradative reactions are
probably photooxidative.  Degradation is due to chemical, rather
than physical change.  This was shown by resolution and casting of
a low elongation film.  A weak, incompatible freshly cast film was
produced.

Table 5.    Accelerated Weathering of Block Copolymer[a]
Films - % Elongation Vs. Hours Exposed -
50 Micron Films - 47% "B" Segment

| HRS. EXPOSED | 0 | 48 | 144 | WET CYCLE |
|---|---|---|---|---|
| POLYMER I | 600 | 280 | 6 | YES |
|  | 600 | 470 | 93 | NO |
| POLYMER II* | 270 | 70 | 6 | YES |
|  | 270 | 175 | 32 | NO |
| POLYMER III | 650 | 325 | 40 | NO |
| POLYMER IV | 330 | 150 | TWTM | NO |

\* "B" CONTAINS $M_N$ 3000 NEOPENTYLGLYCOL ADIPATE

a.    Polymers I and II were prepared using bis(4,4' isocyanato-
cyclohexyl) methane.  Polymer III used 2,4-toluenediiso-
cyanate and Polymer IV used $\alpha,\alpha,\alpha',\alpha'$ - tetramethylxylene diisocyanate

The degradative process causes a marked loss in molecular
weight, Fig. 7.  The diamonds in Fig. 7 show the change in elongation and molecular weight that occurred after the indicated hours
of wet cycle weathering of Polymer I.  The circles are data taken
during the polymerization process described previously in Fig. 3
and Table 2.  A common curve can be drawn through the points, suggesting the loss in elongation upon weathering is due to chain
breaking in the "B" segment to form AB'blocks.  As previously noted
such polymers have poor tensile properties.

Further support Fig. 8, for chain breaking in the polyester/
urethane soft segment is shown by infrared studies of size exclusion
chromatography fractions where fractionation by molecular weight (MW)
shows the low MW material in undegraded polymer is rich in polymethyl
methacrylate (probably non-functional AIBN initiated blocks) while
the high MW material is rich in polyester.  (Urethane extended and
PMMA capped block copolymer).  After 200 hours weathering, the low
MW material is now rich in polyester.  This indicates multiple breaks
have occurred in the polyester urethane segments.

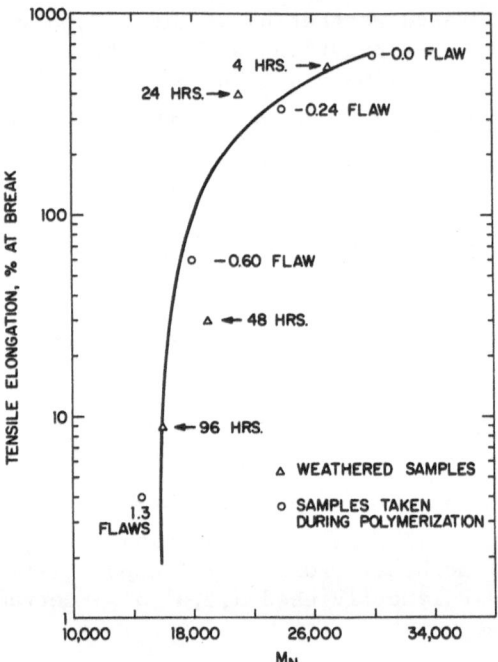

Fig. 7.  Reduction of $M_n$ and Tensile Elongation
         when Polymer I is Subjected to
         Accelerated Weathering.

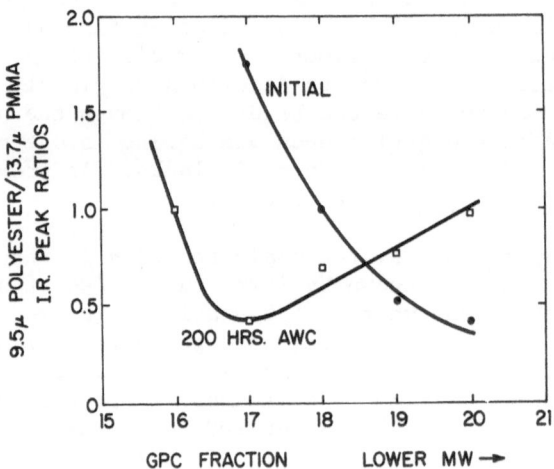

Fig. 8.  Compositional Change in Low Molecular Weight Material
         of Polymer I upon Accelerated Weathering (AWC)

The soft segment contains many potential points for photooxidative attack. Attempts were made to eliminate the most likely possibilities in turn to see if the rate controlling site could be located.

By analogy to polyamide photooxidation, the hydrogen $\alpha$ to the urethane nitrogen was investigated as the point of attack. Polymers having the urethane linkages shown in Fig. 9 were prepared to include the presence (I and II) and absence (III and IV) of hydrogens $\alpha$ to the urethane nitrogen. All the polymers lose elongation at essentially the same rate, Table 5.

POLYMER

I & II

$$-CH_2-O-\overset{O}{\overset{\|}{C}}-\overset{H}{\overset{|}{N}}-\overset{CH_2-CH_2}{\underset{CH_2-CH_2}{CH}} \quad HC\overset{CH_2-CH_2}{\underset{CH_2-CH_2}{CH_2CH}} \quad \overset{CH_2-CH_2}{\underset{CH_2-CH_2}{HC}}\overset{H}{\overset{|}{N}}-\overset{O}{\overset{\|}{C}}-OCH_2-$$

III

$$\text{(aromatic ring with } CH_3 \text{ ortho to } -\overset{H}{\overset{|}{N}}-\overset{O}{\overset{\|}{C}}-OCH_2- \text{ and } NH\overset{O}{\overset{\|}{C}}OCH_2- )$$

IV

$$CH_3-\overset{CH_3}{\underset{|}{C}}-\overset{H}{\overset{|}{N}}-\overset{O}{\overset{\|}{C}}-OCH_2- \quad (\text{benzene ring}) \quad CH_3\overset{}{\underset{CH_3}{\overset{|}{C}}}-\overset{H}{\overset{|}{N}}-\overset{O}{\overset{\|}{C}}-OCH_2-$$

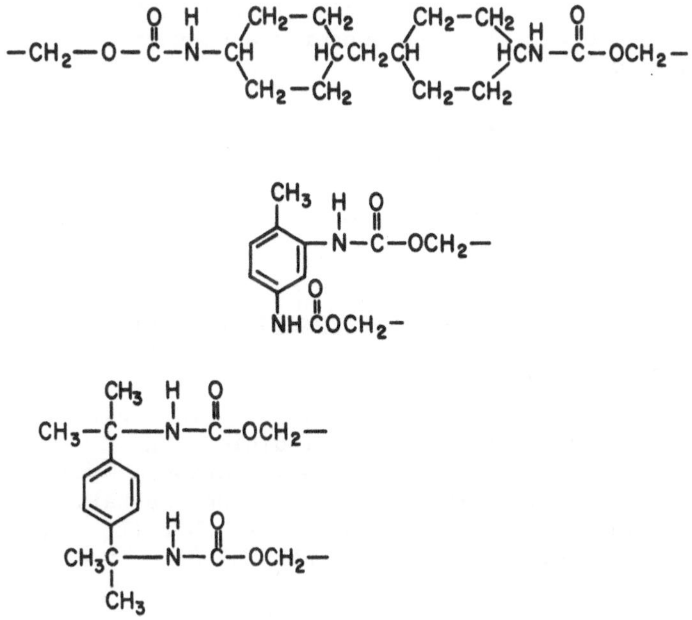

Fig. 9. Connecting Urethane Links in Polymers I, II, III, IV

The ester links were also considered as likely weak points. The lack of benefit from the neopentyl structure, Polymer II, Table 5, or reduced concentration of ester (not illustrated, but tested with a polyester from neopentyl glycol and dodecanedioic acid) indicate this feature is not controlling the degradation process.

The principle remaining feature, the sulfide links between the "A" and "B" segments remains to be eliminated from consideration. Much recent interest by others in the photooxidation of monomeric sulfides supports this possibility (7). The alternative that photooxidative degradation in these polymers is a relatively random process should also be considered.

Polymer Stabilization

Pigments such as channel carbon black and silica coated rutile titanium dioxide retard, Fig. 10, the loss of Polymer I elongation upon accelerated weathering. The trend with pigmentation is as expected and is inversely related to the UV transparency of the film. The weathering of a random high molecular weight polyester urethane mixed with cellulose acetate butyrate (CAB) is also described in Fig. 10. It loses elongation much less rapidly than Polymer I, probably because the fragments formed early in the degradation are not A-B in nature and are still fairly high in molecular weight. The data in Fig. 10 also supports a photooxidative degradation mechanism.

Two UV stabilizers gave the results in Table 6. These materials produced a 2-4X increase in the useful life of the film. The positive effect of these materials on film life also supports a photooxidative mechanism for film failure upon weathering.

CONCLUSIONS

A simple, versatile synthesis of triblock polymers was demonstrated.

Conversion of polymethyl methacrylate polyesterurethane triblock copolymer to a mixture of diblock and triblock by photooxidation rapidly reduces polymer elongation. Over 90% of the elongation is lost when the polymer has sustained breaks equivalent to one in each molecule. This is another illustration of the poor physical properties seen with AB diblock polymers.

Photooxidative chain cleavage occurs in the "B" segment of the ABA block copolymer. By a process of elimination the urethane and

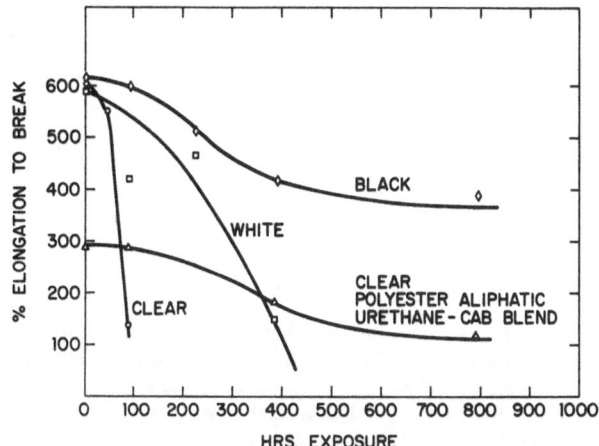

Fig. 10.  Photooxidative Degradation of Polymer I as a
Function of Pigmentation - Accelerated
Weathering Dry Cycle

Table 6.  Evaluation of UV Stabilizers

### STABILIZED POLYMER I
### 50 MICRON FILMS -%ELONGATION
### AS A FUNCTION OF HOURS ACCELERATED AGING - DRY CYCLE

| STABILIZER | 0 | 96 | 200 | 400 | 800 HRS. |
|---|---|---|---|---|---|
| NONE | 600 | 140 | 0 | | |
| 3% n-DODECYL-O-HYDROXYBENZOPHENONE | 750 | 660 | 350 | 0 | |
| 5% ISOAMYLHYDROXY BENZTRIAZOLE | 700 | 680 | 580 | 550 | 0 |

ester groups were determined not to be rate controlling in the degradation process.  Theories supporting attack at sulfide links
between the "A" and "B" blocks or a relatively random attack in the
soft segment  remain to be tested.

Resistance to photooxidative degradation can be improved by
incorporating UV absorbers or opaque pigments into the polymer
film.

REFERENCES

1.  D. C. Allport and W. H. Janes, "Block Copolymers", John
Wiley & Sons, NY-Toronto, a Halsted Press Book, 1973.
2.  G. E. Molar, "Colloidal and Morphological Behavior of
Block and Graft Copolymers," Plenum Press, New York, London, 1971.
3.  J. H. Grezlak and G. L. Wilkes, J. of App. Pol. Sci., 19,
769-786 (1975).
4.  A. Noshay, M. Matzneu and L. M. Robeson, J. Pol. Sci.,
Polymer Symposia 60, 87-95 (1977).
5.  A. C. Soldatos and A. S. Burnhans, Ind. Eng. Chem.
Product Res. Dev., 9, 296 (1970).
6.  R. F. Fedors, Polymer 20, 518-519 (1979).
7.  D. Sinnreich, H. Lind and H. Batzer, Tetrahedron Letters
39, 3541-3542 (1976).

R. Ranby and J. F. Rabek, "Singlet Oxygen, Reaction with
Organic Compounds and Polymers",Chapt. 29, p. 288. Pub. by John
Wiley & Sons, Chichester, New York, Brisbane, Toronto, (1978).
M. L. Kacher, "Mechanisms of Sulfide Photooxidation,"
Ph.D. Thesis, U. of California, Los Angeles, (1977), Xerox Univer-
sity Microfilms, Ann Arbor, Michigan, 48016, Items #77-30,927.

INTERPENETRATION OF SILICA IN A NETWORK OF CELLULOSE AND DIVALENT

LEAD TO FORM GLASSY POLYMERS

T. L. Ward, W. R. Goynes, Jr., and R. R. Benerito

Southern Regional Research Center
Science and Education Administration
U.S. Department of Agriculture
New Orleans, Louisiana 70179

INTRODUCTION

Glassy polymeric materials were prepared by heating sodium plumbite-treated cotton in contact with a siliceous substrate. The glassy polymeric materials resemble petrified cotton and contain considerable carbon. They resist acid and alkali, are hard, and can serve as labels, decorations, adhesives between glasses, or colored coatings.

These new materials seem to exhibit characteristics of both polymers and glasses. Usually a "glass" contains carbon only in trace amounts as a coloring agent or as a means of producing foamed glass, whereas polymers often contain carbon compounds. Glasses contain silicon and oxygen, but so do silicone polymers. Both polymers and glasses are composed of repeating units.

Lead oxide has been widely used in the glazing of pottery, in flint-type glasses, and in solder glasses, but these materials contain no carbon and have no visible structure. In the process being reported, carbon contained in the cotton is required for formation of the new chemical structure and for retention of the fabric weave pattern and shape in the new products.

The novel mechanism by which these glassy materials are formed involves movement of silicon from the substrate across the interface to the carbonaceous skeleton of the treated cotton. The migration in solid states occurs at temperatures considerably below the melting point of either the siliceous substrate or polymeric plumbous oxide. The relatively low temperature of the process suggests that interpenetration by the silicon occurs by an interstitial mechanism.

EXPERIMENTAL

The procedure for treating the fabric used in making the glassy polymer has been described previously (1). It consists of immersing cotton fabric in an aqueous solution of sodium plumbite, washing out the excess reactant with water, and drying the fabric. Lead content of the fabric can be varied up to a maximum of about 50% (by weight) by varying immersion time from about 1 to 24 h. The fabric may be of any construction, such as plain weave, knit, or lace.

The plumbite-treated fabric, cut to desired shape, is heated in a closed oven to approximately 600°C, with the fabric in surface contact with either a silicate glass or porcelain. It is important that the oven remain closed while the temperature is raised to about 300°C to prevent ignition of the fabric. Above 300°C, the oven can be opened periodically for examination of the sample. The temperature of 600°C is maintained for about 1 h. The maximum temperature required will vary somewhat depending on the siliceous substrate used. Temperatures as low as 550°C with soda-lime glass and as high as 700°C with borosilicate glasses have been used with success.

Several techniques have been used to study the new glasses and the mechanism for their production. Lead contents of the treated fabrics were determined by X-ray fluorescence with disks of ground material. Properties of glassy polymers were determined by electron spectroscopy for chemical analyses (ESCA), scanning electron microscopy (SEM), optical microscopy, and energy-dispersive X-ray analysis (EDAX) by use of equipment and techniques described previously (2). Thermogravimetric analysis (TGA) and differential thermal analysis (DTA) were performed with a Dupont 900 analyzer equipped with high temperature modules.[*] The glassy polymers were immersed in a strong acidic or basic solution for 24 h to extract loosely held lead. Lead contents were determined by atomic absorption (2).

F-centers as a source of color were investigated by irradiating the glassy polymers with cobalt-60 to a dosage of 1 Mrd or with ultraviolet at 3500 Å and 2537 Å at an intensity of 9200 $\mu$ w/cm$^2$. Length of exposure varied from 10 min to 6 h.

RESULTS AND DISCUSSION

Cotton fabrics with lead contents varying from 4 to 50% by weight were placed on porcelain, borosilicate glass, soda-lime

*Names of companies or commercial products are given solely for the purpose of providing specific information; their mention does not imply recommendation or endorsement by the U.S. Department of Agriculture over others not mentioned.

glass, or other hard silicate substrates and heated as previously
described (3).  All of the samples formed glassy polymeric materials.
Fabrics containing from 6 to 15% lead formed translucent films that
adhered firmly to the substrate surface on which they were produced.
These films had the same shape and weave characteristics as the
treated fabric from which they were made.  In Figure 1 are shown the
glassy coatings formed when a treated cotton cheesecloth is reacted
with three types of substrates:  a borosilicate flask, a piece of
Fibrex glass, and a soft glass vial.  These coatings made from a
fabric with a low lead content can be written on with an ordinary
lead pencil and so can serve as labels.  They can also serve as
decoration.  Cotton lace is particularly attractive when glassified
in this manner.  Thick fabrics and multilayers of a given fabric
have been used, but in all cases, the reaction occurs at the solid-
solid interface only.  Thick coatings cannot be formed by using
multilayers of treated fabrics.

Fig. 1.  Cotton cheesecloth (lead content 8%) on left to right:
         borosilicate flask, Fibrex glass slab, soft glass vial.

Glassy polymers from fabrics with more than 15% lead separate from the substrate, usually as brittle flakes. When these brittle flakes are removed, an etched replica of the fabric shape and weave pattern remains on the substrate. This etched substrate surface shows no lead when examined by ESCA. Thus, there is only diffusion of silica from the substrate to the lead-carbon cellulosic matrix without reciprocal diffusion of lead into the original silica substrate. The higher the lead content of the fabric, the smoother is the surface of the new glassy material. However, the fabric weave pattern is still visible in all glassy products.

Although cotton in fabric form was usually used, other cellulosics performed well. The graph in Figure 2 shows that cellulose as a cotton fabric, paper sheet, or wood veneer picked up similar amounts of lead from the plumbite solution. All were successfully converted into the glassy state, although wood tended to wrinkle and yield an uneven product. Because paper and wood veneer have no weave or open structure, the resultant glassy materials were smooth.

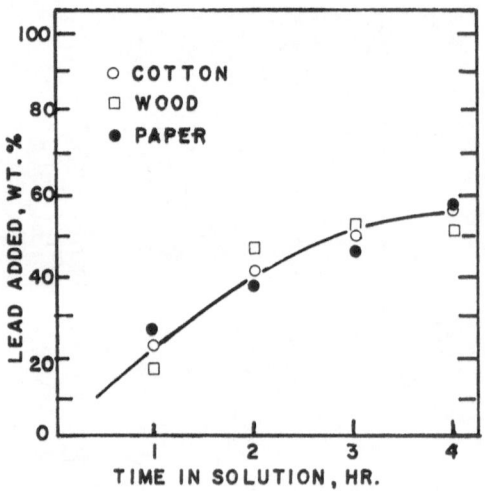

Fig. 2.    Cotton printcloth, pine wood veneer, or filter paper
          immersed in sodium plumbite (12% aqueous base) at 26°C,
          washed, and dried.

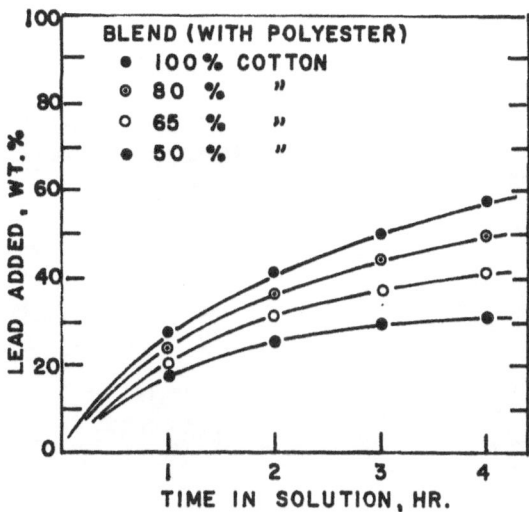

Fig. 3.   Pure cotton and intimate blends of cotton with polyester
          as twill weave, immersed in sodium plumbite solution (12%
          aqueous base) at 26°C, washed, and dried.

     The curves in Figure 3 show that after several hours, blends
of cotton with polyester picked up lead from sodium plumbite solution in amounts that varied linearly with the percentage of cellulose.  Although blends with as little as 50% cellulose formed glassy
materials, fabrics containing at least 80% cellulose preserved the
weave pattern better in the glassy product.

     The strongly adherent polymers made from fabrics containing
less than 15% lead were used to join borosilicate, quartz, porcelain, and soda-lime glasses.  Various combinations were joined by
placing a piece of plumbite-treated cotton fabric between the
pieces and heating the assembly in the furnace to 600°C.  A quartz
cuvette bonded to a borosilicate plate by this process is shown in
Figure 4.  After cooling, the bond was so strong that the joined
glasses ruptured more readily than the bond.  The fabric can be
ground and mixed with glycerol to form a paste that can be substituted for the fabric in bonding various glasses.

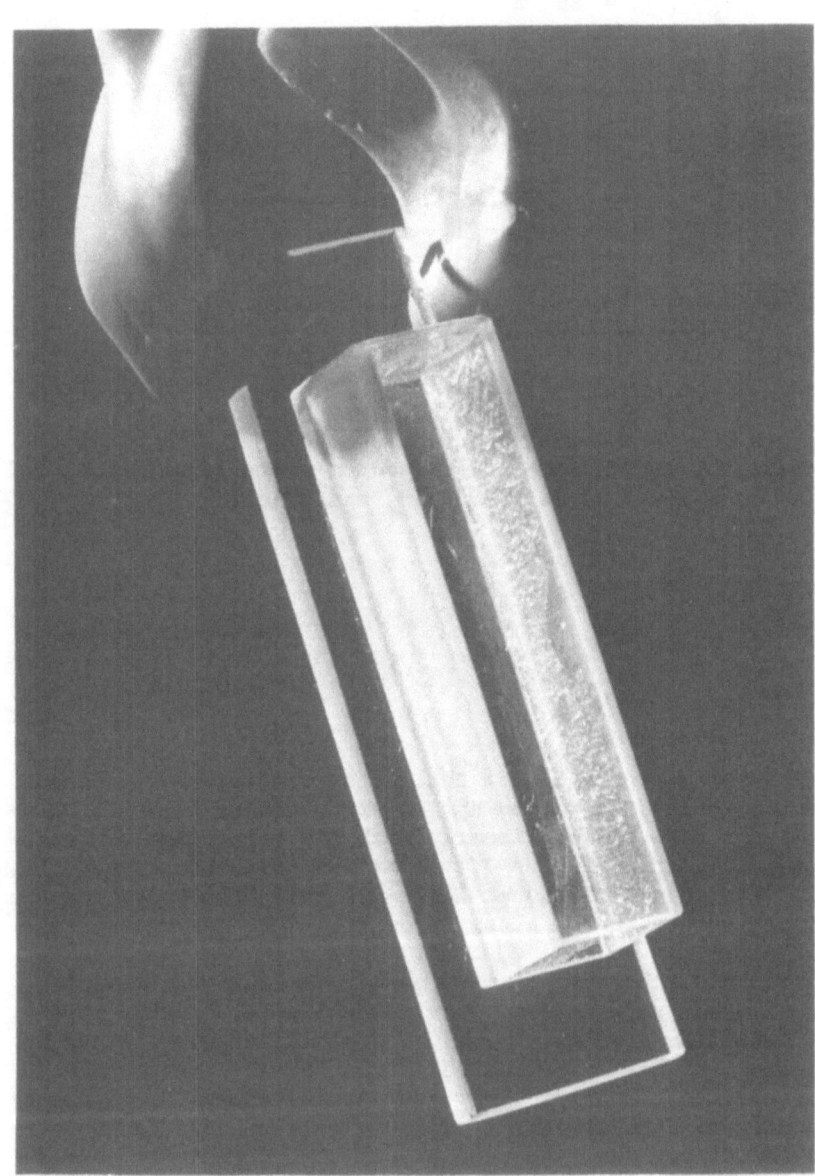

Fig. 4.  Quartz cuvette on borosilicate plate, Cotton printcloth (12% lead) bonding agent, Assembly heated to 600°C and held 1 h.

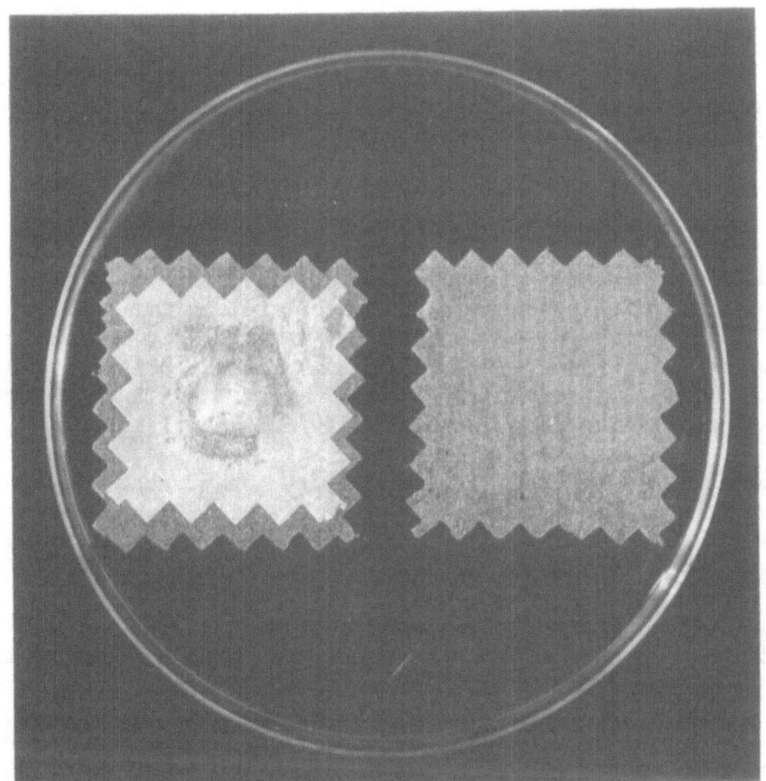

Fig. 5.   Left:   Cotton 80 X 80 printcloth (8% lead) with
          aluminum foil on borosilicate glass.  Right:
          Control cotton without foil.  Assembly heated
          to 600°C and held 1 h.

The appearance and electrical properties of the glassy
materials were altered by heating either aluminum foil or powder
spread evenly on top of the treated fabric.  The left side of
Figure 5 shows how prolonged heating at 600°C results in disap-
pearance of the foil or powder into the film.  Before complete
absorption, the surface is semiopaque, gray, and electrically con-
ductive.  Prolonged heating at 600°C also lowered the electrical
conductivity.

Other elements such as tin, copper, barium, bismuth, hafnium,
zirconium, tungsten, and molybdenum were applied to cotton in a
manner analogous to that used with lead.  No glass formed from cottons treated solely with salts of the other metals.  Molybdenum, in
concert with lead, did result in a glassy product, but the product
was inferior to the one produced by lead alone.

All high-silicate glasses served well as the substrate for the
process although borosilicate glass performed best. This result is
partly because the softening point of borosilicate glass is sufficiently higher than 600°, and thus less care is required to prevent
overheating. Also, the higher resistance of borosilicate glasses
to thermal shock allows for more rapid heating and cooling of
assemblies.

Other siliceous materials such as sand, silicon dioxide,
sintered silicon, silicic acid, water glass, calcium-silicate, and
silicone vacuum grease were tried as the substrate material. Only
the silicone grease acted as a source of silicon for this process,
and the product was white, opaque, and fragile, resembling eggshell.
With the silicone grease, the fabric shape was retained but not the
weave pattern. Silicon-to-oxygen bonding in a silicone resembles
that of a glass more than that of the other materials tested.

The glassy polymers are hard and are not scratched by soft
glass or by stainless steel. After soaking for 24 h in concentrated
$HCl$, $HNO_3$, $H_2SO_4$, aqua regia, or 50% NaOH in water, the solutions
contained less than 20 ppm of lead.

During the heating, the plumbite-treated fabric turns black at
about 250°C. In this state, the fabric has enough strength that it
can be moved about. Further heating to 400°C turns the fabric
yellow. This material is too fragile to be moved even though it
appears to be yellow fabric. When the temperature is being raised
from 500°C to 600°C, the oven can be opened every 10 to 20 degrees
and the changes observed. In this temperature range, the fragile
yellow fabric appears to be pulled tightly to the surface. It even
follows compound curvatures of the substrate. It appears almost as
if the substrate exerts a magnetic attraction on the yellow fabric
material. Continued heating to 600°C causes the fabric skeleton to
become glassy.

X-ray diffraction patterns showed similar strong bands for
charred treated fabric heated to 500°C in the oven and untreated
fabric mixed with lead monoxide. This result indicates that the
charred, leaded cotton contains lead monoxide. Prior to any heating, the treated fabric gave diffraction peaks characteristic of
cellulose I and three new peaks that differed from those of lead
monoxide.

X-ray diffractograms of the glassy polymers did not give sharp
peaks but showed amorphous halos that indicate no long range order.
This type of pattern is characteristic of random network polymers.
As the lead content of the fabric was increased, the "d" spacing
of the broad halo decreased for the glassy polymer made from it.
The glassy polymers did not show the $SiO_2$ band present in the
siliceous substrate.

As the fabric is heated, it shrinks in all directions uniformly. This shrinkage causes the lead content of this heated material to be about 20% higher than the fabric prior to heating. EDAX analysis (Figure 6) revealed even distribution of lead throughout cross sections of both the unheated control and the skeleton of the heated fabric. EDAX analyses also showed that silicon was evenly distributed throughout the skeleton formed when the sample was heated while in contact with the siliceous substrate, and that its concentration increased with increasing temperature.

Photographs taken through an optical microscope of front-lit glassy materials made from cotton fabrics of low and high lead contents are shown in Figure 7. The surface of the material made from

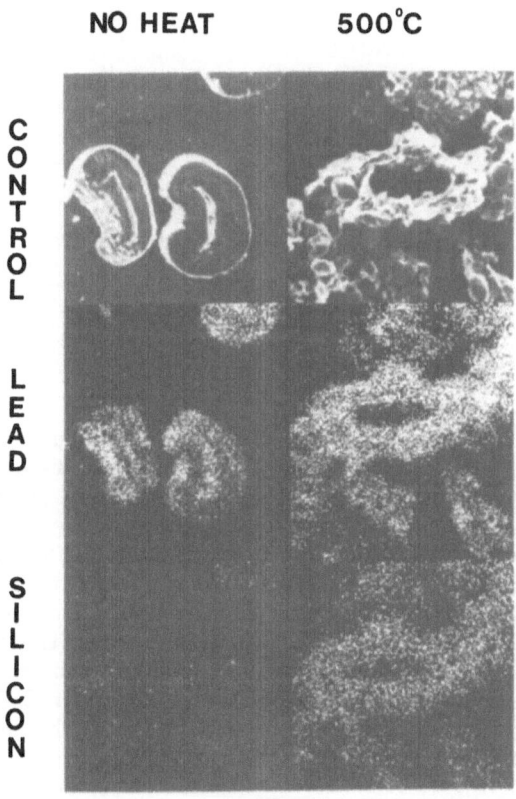

Fig. 6.    Cross sections of fibers from plumbite-treated
           80 X 80 cotton printcloth (16.87% lead), unheated
           and heated in contact with borosilicate glass.
           Examined by EDAX for lead and silicon.

A

B

Fig. 7,  Front-lit optical microscope photos (50 X).
        A, Glassy material from low-lead fabric (8%)
        B, Glassy material from high-lead fabric (20%)

the low lead fabric is rough and fibrous and the weave pattern is
easily discernible in the photograph.  The texture explains why the
surfaces of products from fabrics of low lead content can be written
on with an ordinary lead pencil.  When the starting fabric is of a
high lead content, the product has a smoother surface, which shows
in the enlargement of surface features.  The parallel lines of the
weave pattern can still be seen in the photograph and are even more
discernible with the naked eye.

     Scanning electron micrographs of the skeleton of the sample
heated to 400°C (Figure 8) revealed a structure at low magnification
similar to that of the unheated fabric.  In the sample heated to
500°C, there was some shrinkage of fibers with a consequent enlargement of weave openings and a small amount of fiber breakage.

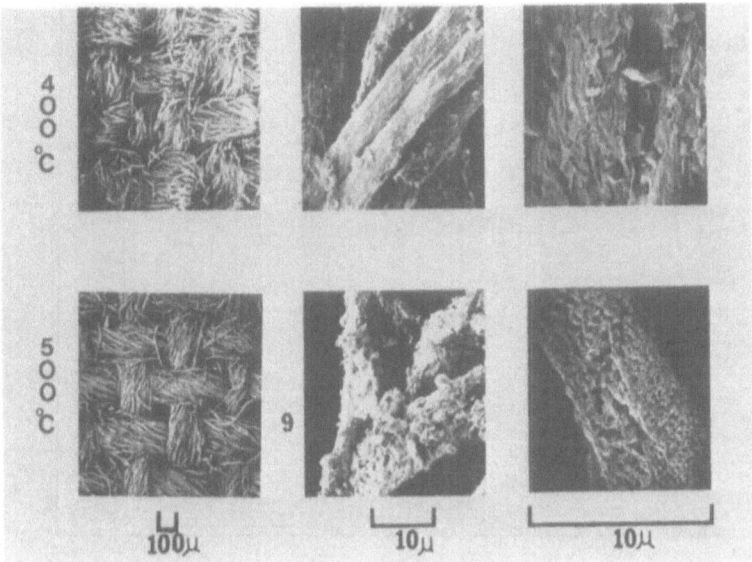

Fig. 8.   Cotton 80 X 80 printcloth (16.27% lead).

At high magnifications changes that occur at about 500°C become
evident.  The fibrous character is replaced by a fused, crystalline
skeleton.  This skeleton serves as the pattern from which a
replicate of the starting fabric is made.

Data obtained by ESCA of glassy polymers made from cotton
fabrics with different levels of lead content indicated that the
atom ratios of silicon to oxygen and of lead to carbon are similar
in the different products.  There is 1 lead per 5 carbon atoms and
10 oxygen atoms per silicon atom.  These data suggest that similar
bonds between silicon and oxygen and between lead and carbon are
formed in the different products, and indicate chemical bonding
rather than deposition.

The atom ratio of carbon to silicon in glassy material made
from low-lead fabric is greater than that made from a high-lead
fabric.  This finding explains the more obvious weave pattern and
translucent appearance of materials made from a fabric of low lead
content.  Prior to heating, a fabric with 8.2% lead has 1 lead
atom per 90 carbon atoms.  After conversion to the glassy state, it
has 1 lead atom per 5 carbon atoms; it retains one-eighteenth of the
original carbon.  By contrast, fabric with 37% lead retains two-
thirds of its carbon.  Increased lead content in the fabric

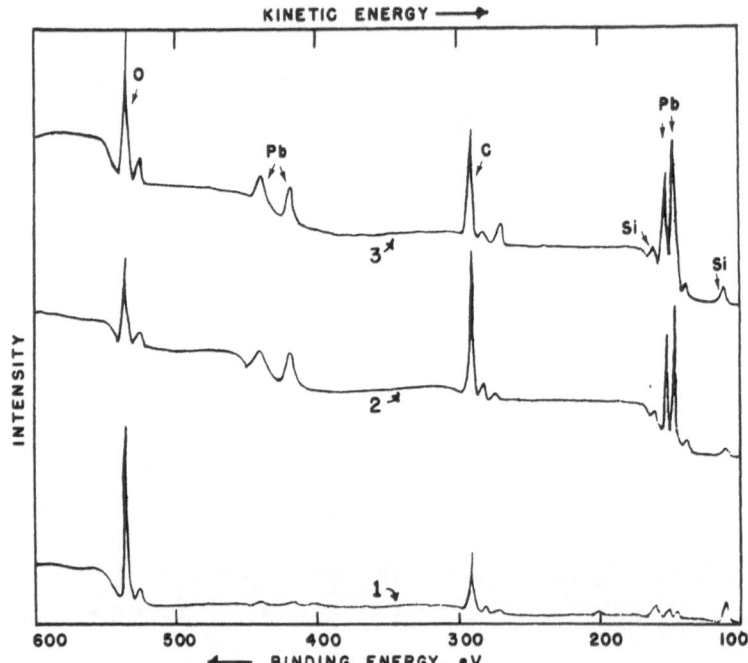

Fig. 9.   ESCA spectra for:  1, borosilicate substrate; 2,
          glassy material from cotton fabric (16.87% lead)
          converted to glassy state; 3), material of 2 turned
          to silver by flame.

increases the ability of silicate to migrate across the interface
between the fabric and substrate.

    When treated fabric and borosilicate glass were heated in a
TGA-DTA instrument in 10% oxygen in nitrogen, all of the weight
loss occurred before the temperature reached 525°C.  Differential
thermal analysis showed two exothermic events, both occurring prior
to 525°C.  No change occurred in either TGA or DTA curves as the
materials were further heated to 1200°C in an atmosphere of 10%
oxygen with nitrogen.  Similar analysis in pure nitrogen resulted
in production of a fibrous char.  Analyses performed in the atmosphere of 10% oxygen produced the glassy product that is produced
in the usually used oven.  The newly formed glassy polymers are
stable in the Analyzer to 1200°C, the upper temperature limit of
the equipment.  A comparison of temperatures at which visible
changes occurred in the oven with those of significance in the
TGA-DTA analysis, led to the conclusion that the glassy materials
are formed just above 500°C.

The new glassy polymers are colorless when formed. Exposure to the direct flame of a bunsen-type burner induces a gold, purple, or silver-gray color that can be removed by reheating the sample to 600°C in the oven or by applying a flame to the reverse side of the glassy material. This color change is reversible for at least four cycles, the maximum number tried. ESCA and electrical conductivity measurements showed that the color is not due to reduction of lead to the metallic form. ESCA spectra (Figure 9) show that the oxidation state of lead in the colored coating that had been turned silver by a flame and in the material prior to flame treatment is identical. The color change could not be induced by irradiation with cobalt 60 for 3 h or ultraviolet at 3500 Å or 2537 Å for 6 h. This result indicates that the color change was not because of an F-center effect. The color change may result from either changes in particle sizes or in ordering of the lattice structure.

CONCLUSION

The results indicate a novel glassy material and process for its production. Because the carbon-to-lead atom ratio of the glassy materials is about 6:1 and because lead is not known to form a carbide, the mechanism may be an example of diffusion by an interstitial mechanism. When the cellulosic fabric containing lead is heated to 500°C, the skeleton is composed of lead, carbon, and oxygen having the same weave pattern and shape as the starting material. Because the atom ratio of carbon to lead in all final products is six, the excess carbon must be driven off. Continued heating permits silicon to migrate from the substrate into the skeleton to form new bonds, and the amount of silicon attracted is proportional to the amount of lead present in the starting cellulosic material. The more lead that is present, the more silicate that is removed from the substrate surface and the more likely is a fracture at the interface upon cooling. Also, because a smaller proportion of silicon atoms is taken into the skeletal matrix when the lead content is low, the resultant product has more lead-oxygen-carbon skeleton retained as weave pattern or fabric design. F. L. Vogel and J. N. Zemel of the University of Pennsylvania have observed that hexagonal layered structures form when cellulosic materials are charred and that the relatively low temperature of our process suggests diffusion by an interstitial mechanism (4). It would seem that the order of hexagonal layered structure would show in X-ray diffraction and other studies that showed no long-range order for our materials. The lack of long-range order imparts glasslike properties and poor electrical conductivity except when metallic aluminum is partially exposed in the surface.

This study has shown that diffusion of silicon into the lead-treated cotton occurs at reasonably low temperatures. It is possible that such diffusion can be used to preserve documents and to produce objects of art or decoration that are not presently available.

ACKNOWLEDGMENTS

The authors thank J. Carra, J. Harris, J. Hebert, D. Mitcham, L. Muller, B. Piccolo, and D. Soignet for analyses.

REFERENCES

1.  T. L. Ward and R. R. Benerito, "Addition of Lead from Sodium Plumbite Solution to Modified and Unmodified Cottons," Text. Res. J. 44, 12 (1974).
2.  R. T. O'Conner, "Instrumental Analysis of Cotton Cellulose and Modified Cotton Cellulose," Marcel Dekker, Inc., New York, 1972.
3.  T. L. Ward and R. R. Benerito, "Coatings Formed at the Interface of Glass and Plumbite-Treated Cotton," Thin Solid Films 53, 73-79 (1978).
4.  F. L. Vogel and J. N. Zemel, "Petrified Cotton?" Nature (London) 275, 11 (1978).

ASPECTS OF RELATIVE NETWORK CONTINUITY AND PHYSICAL CROSSLINKS VIA
AN ANALYSIS OF POLYSTYRENE/POLYSTYRENE HOMO-INTERPENETRATING POLYMER
NETWORK LITERATURE

D. L. Siegfried, D. A. Thomas, and L. H. Sperling

Materials Research Center, Coxe Lab #32
Lehigh University
Bethlehem, Pa.  18015

ABSTRACT

   Sequential interpenetrating polymer networks, IPN's, are
synthesized by swelling a crosslinked polymer I with monomer II, plus
crosslinking and activating agents, and polymerizing II in situ.
The materials are called homo-IPN's if polymers I and II are
chemically identical.

   Because of the special swelling and mutual dilution effects
encountered in sequential IPN's, special equations were derived
for their rubbery modulus and equilibrium swelling.  The new
equations were used to analyze polystyrene/polystyrene homo-IPN
swelling and rubbery modulus data obtained by four different
laboratories.  In the fully swollen state, there was no evidence
for IPN related physical crosslinks, but some data supported the
concept of network I domination.  In the bulk state, network I
clearly dominates network II because of its greater continuity in
space.  The analysis of the data concerning the possible presence
of added physical crosslinks in the bulk state yielded inconclusive
results, but this latter is of special interest for modern network
theories.

ACKNOWLEDGEMENT

The authors wish to acknowledge the support of the National Science
Foundation through Grant No. DMR77-15439 A01, Polymers Program.

INTRODUCTION

In recent years, important advances in the theory of rubber elasticity have been made. These include the introduction of the so-called phantom networks by Flory (1) and a two-network model for crosslinks and trapped entanglements by Ferry and coworkers (2,3). The latter builds on work by Flory (4) and others on networks cross-linked twice, once in the relaxed state, and then again in the strained state. In another study, Kramer (5) and Graessley (6) distinguished among the three kinds of physical entanglements as crosslink sites, the Bueche-Mullins trap, the Ferry trap, and the Langley trap.

These papers discuss the contribution, or lack of contribution, of physical crosslinks to the theory of rubber elasticity (7). Through the use of sequentially polymerized homo-interpenetrating polymer networks, IPN's, a method will be described below to detect changes, if any, in the physical crosslink density, provided that such physical crosslinks actually do contribute to the retractive and swelling forces.

Sequential interpenetrating polymer networks, are synthesized by swelling a crosslinked polymer I with monomer II, plus crosslinking and activating agents, and polymerizing monomer II in situ (8-13). In the limiting case of compatibility between crosslinked polymers I and II, both networks may be visualized as being interpenetrating on a molecular level and continuous throughout the macroscopic sample.

If both polymer networks are chemically identical, the naive viewpoint suggests that a mutual solution will result with few ways to distinguish network I from network II. As will be shown below, this is emphatically not the case.

There have been four major research reports using IPN's where both networks are identical (14-17) as well as other studies and applications (18-20). All four major studies, the basis for this analysis, interestingly enough used networks of polystyrene (PS) crosslinked with divinyl benzene (DVB). The first publication on PS/PS-type IPN's was a study of swelling by Millar which appeared in 1960 (15). [IPN's having both networks identical in chemical composition have sometimes been called Millar IPN's, after his pioneering work (14)]. Shibayama and Suzuki published a paper in 1966 (16) on the modulus and swelling properties of PS/PS IPN's, followed by Siegfried, Manson and Sperling (14), who also examined viscoelastic behavior and morphology. Most recently, Thiele and Cohen (17) studied swelling and modulus behavior, and derived a key equation with which to study the swelling behavior of IPN's.

The purpose of this paper is to reexamine the studies on PS/PS

IPN's and scrutinize the results in the light of new theoretical developments (1-6). For clarity, all PS/PS IPN's will be denoted by two pairs of numbers: actual volume % DVB in polymer I/actual volume % DVB in polymer II, % polymer I/% polymer II.

## SYNTHETIC VARIATIONS

While all four investigators empolyed PS/PS IPN's crosslinked with DVB, each differed in important details. Table I summarizes the principal variations (14-17). The extent of swelling and polymerization conditions probably influence the final results.

## DEVELOPMENT OF THEORY

In order to properly analyze and compare the data in the four papers, two equations especially designed for compatible IPN's were derived. These equations relate the swelling and modulus behavior to the double network composition and crosslink level. In all cases, a sequential mode of synthesis is assumed where network II swells network I.

### Sequential IPN Rubbery Modulus

Let us consider the Young's modulus, E, of a sequential IPN having both polymers above their respective glass transition temperatures. A simple numerical average of the two network properties results in (21):

$$E = 3(N_1 v_1 + N_2 v_2) RT \tag{1}$$

where $N_1$ and $N_2$ represent the number of moles of network I and II chains per $cm^3$, respectively, and $v_1$ and $v_2$ are the volume fractions of the two polymers, respectively. The quantities R and T stand for the gas constant and the absolute temperature, respectively. While equation (1) ought to describe simultaneous interpenetrating networks, SIN's (22,23) interestingly enough, it has never been tested. Nevertheless, it does not adequately express the Young's modulus of a sequential IPN.

An equation for a sequential IPN begins with a consideration of the front factor, $\overline{r_i^2} / \overline{r_f^2}$ (24,25), where $r_i$ represents the actual end-to-end distance of a chain segment between crosslink sites in the network, and $r_f$ represents the equivalent free chain end-to-end distance. For a single network,

$$E = 3 \frac{\overline{r_i^2}}{\overline{r_f^2}} NRT \tag{2}$$

Table I.  Synthetic Details for DVB/DVB PS/PS IPN'S

| Investigator | Methods | Comments |
|---|---|---|
| Millar | Suspension | Each sample swelled to equilibrium. |
| Shibayama and Suzuki | Bulk | Each sample swelled to equilibrium. |
| Siegfried, Manson and Sperling | Bulk | Controlled degrees of Swelling. |
| Thiele and Cohen | Bulk | Swelled to equilibrium, polymerization in the presence of excess monomer. |

If the network is unperturbed, the front factor is assumed to equal
unity.  For the case of perturbation via network swelling,

$$\frac{\overline{r_i^2}}{\overline{r_f^2}} = \frac{1}{v^{2/3}} \tag{3}$$

For network I swollen with network II, the concentration of $N_1$
chains is reduced to $N_1 v_1$.  Substituting this and equation (3) into
equation (2), the contribution to the modulus by network I, $E_1$,
becomes

$$E_1 = 3\, v_1^{1/3}\, N_1 RT \tag{4}$$

With dilution of network II by network I, the contribution to the
modulus by network II may be written,

$$E_2 = 3\, v_2\, N_2 RT \tag{5}$$

Equation (5) assumes that network II is dispersed in network I, yet
retains sufficient continuity to contribute to the modulus of a
diluted material.  Since the chain conformation of network II
undergoes minimal perturbation, the quantity $\overline{r_i^2} / \overline{r_f^2}$ is further
assumed to be unity.

The modulus contributions may be added:

$$E_1 + E_2 = E \tag{6}$$

This assumes mutual network dilution and co-continuity, with no
added internetwork physical crosslinks, and since the final
material has only the two networks,

$$v_1 + v_2 = 1 \tag{7}$$

Then E may be expressed,

$$E = 3(v_1^{1/3}\, N_1 + v_2\, N_2)\, RT \tag{8}$$

as the final result.  Equation (8) will always yield a larger value
of E than equation (1), because $v_1$ is a fractional quantity.  Further,
equation (8) does not yield a simple numerical average of the two
crosslink densities.

It is of special interest to note that Meissner and Klier
developed an equation to express the behavior of supercoiled

networks, prepared by crosslinking in solution, and evaporation of solvent (26).

## A Modified Thiele-Cohen Swelling Equation

For many years, the Flory-Rehner (27) equilibrium swelling equation has served to characterize single network properties. Recently, Thiele and Cohen (7) derived the corresponding equation for homo-IPN's, i.e., where networks I and II are chemically identical except for crosslink level. Neither the Flory-Rehner equation nor the Thiele-Cohen equation contains the thermoelastic front factor to account for internal energy changes on swelling, although simple analogy with equation (2) shows that one must exist (28). The final equation of state for swelling a sequential IPN reads:

$$\ln(1-v_1-v_2) + v_1 + v_2 + \chi_s \, (v_1 + v_2)^2 =$$
$$-V_s N_1{}' \, (1/v_1{}^o)^{2/3} \, (v_1{}^{1/3} - v_1/2) \qquad (9)$$
$$-V_s N_2{}' \, (v_2{}^{o2/3} \, v_2{}^{1/3} - v_2/2)$$

where the subscript s refers to solvent, $v_1$ and $v_2$ represent the volume fractions of the two polymers in the equilibrium swollen state, $v_1{}^o$ and $v_2{}^o$ represent the volume fractions of the two polymers in the dry state, $\chi_s$ denotes the polymer-solvent interaction parameter (presumed identical for both polymers), $V_s$ represents the solvent molar volume, and $N_1{}'$ and $N_2{}'$ represent the crosslinks of the corresponding single networks in moles/cm$^3$ as determined by the Flory-Rehner equation. Equation (9) differs from the Thiele-Cohen equation by the insertion of the term $(1/v_1{}^o)^{2/3}$ in the first term on the right. Typical values of the term $(1/v_1{}^o)^{2/3}$ range from 1.1 to 4.6 as $v_1{}^o$ varies from 0.9 to 0.1, respectively. The experimental quantity of interest is the total volume fraction of polymer, v, in the swollen IPN:

$$v = v_1 + v_2 \qquad (10)$$

A word must be said about the use of equations (8) and (9). In each case, the modulus of the single networks determines $N_1$ and $N_2$ for equation (8) and the Flory-Rehner equation determines $N_1{}'$ and $N_2{}'$ for equation (9) from the equivalent single network. In this manner, several effects existing in the single networks are minimized, such as physical crosslinks, incomplete crosslinking, and branching. Thus, differences from theory will emphasize new effects due to sequential IPN formation.

## Domain Formation and Internetwork Coupling

Equation (9) is predicated on the assumptions that both networks are continuous in space, network II swells network I, and that the swelling agent then swells both networks. The several PS/PS IPN's under consideration constitute an excellent model system with which to examine fundamental polymer parameters. Questions of interest in the field of IPN's relate to the relative continuity of networks I and II and their consequent relative contribution to physical properties and the extent of formation of physical crosslinks or actual chemical bonds between the two networks. The reader will note that if equation (9) is obeyed exactly, the implicit assumptions require that both networks be mutually dissolved in one another and yet remain chemically independent. Then the only features of importance are the crosslink densities and the proportions of each network.

## RESULTS

### Swelling Data

Figure 1 shows values of N(swelling),(y-axis) determined via the Flory-Rehner equation vs N(stoichiometry), (x-axis) determined by the average quantity of DVB in the IPN. While the IPN's are characterized by two pairs of numbers in Fig. 1, the single networks shown are identified by a single number representing the quantity of DVB used. The fit for the IPN's is extremely poor. (For convenience, log-log plots are presented, the scatter for the linear plots yields the correct perspective.) While some of the data lies below the theoretical line, some lies well above it.

Values for v obtained via equation (9) vs v's determined from swelling the IPN's in toluene are shown in Figures 2 and 3 for the Thiele-Cohen equation and the modified Thiele-Cohen equation, respectively. While both equations fit the IPN data much better than the original Flory-Rehner equation, see Figure 1, the present modification fits somewhat better, on the average.

If new physical or chemical crosslinks were added during IPN formation, one would expect the data to be shifted to the right of the theoretical line. Figure 3 indicates that substantially no new physical or chemical crosslinks are present, at least when the networks are fully swollen. However, analysis of the data does indicate that the contribution to the swelling by network I is out of proportion to its crosslink density and proportion in the IPN. The clearest indication of the conclusion can be found by examining the 4/1, 4/4, and 4/10 crosslink levels of the 44/56 composition samples investigated by Millar (15). Note that the experimental v's (x-axis) substantially do not vary in this series, while both the Thiele-Cohen (17) and modified Thiele-Cohen equations predict

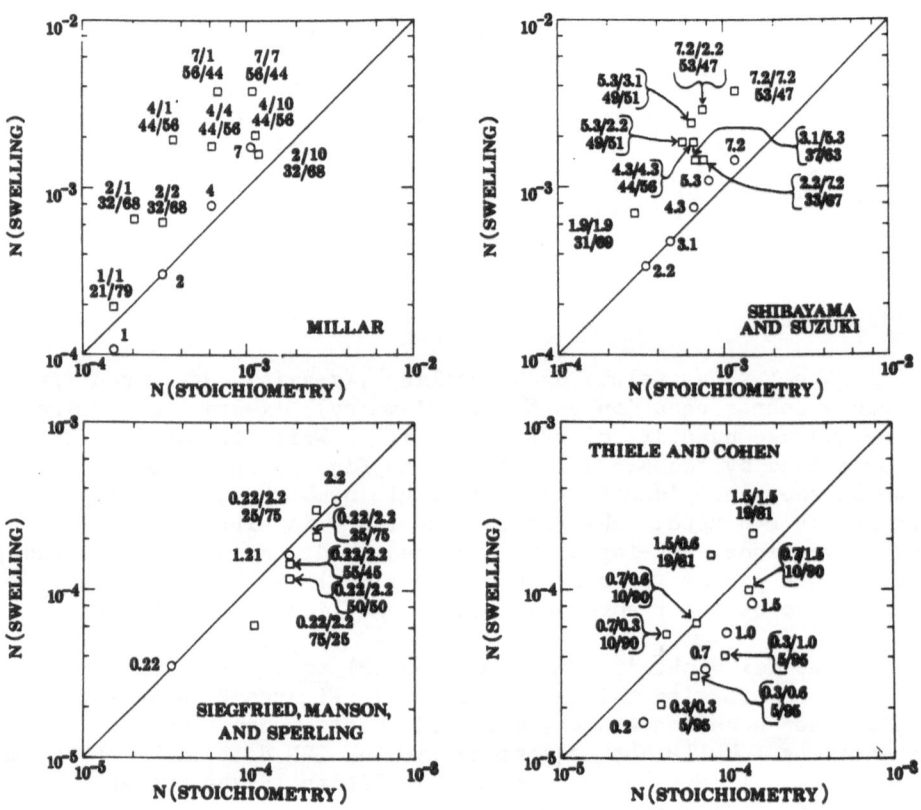

Fig. 1.   Crosslink densities via the Flory-Rehner equation vs
          stoichiometric crosslink densities.   Squares represent
          PS/PS IPN's and circles represent PS single networks.

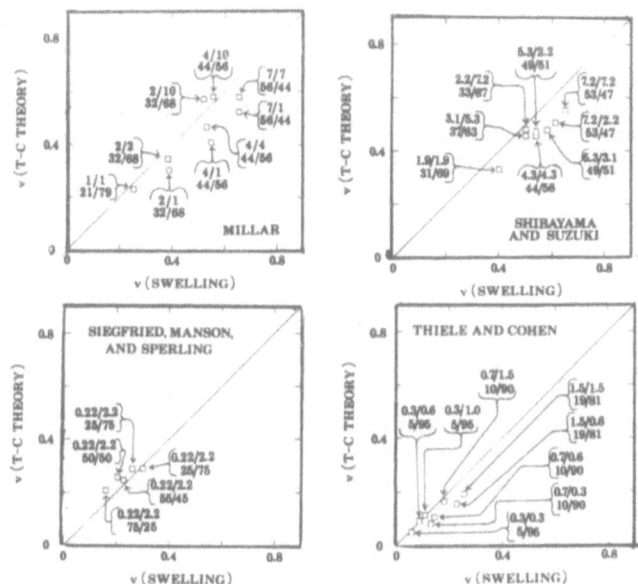

Fig. 2.   Swelling values predicted from the Thiele-Cohen
equation <u>vs</u> experimental swelling values.

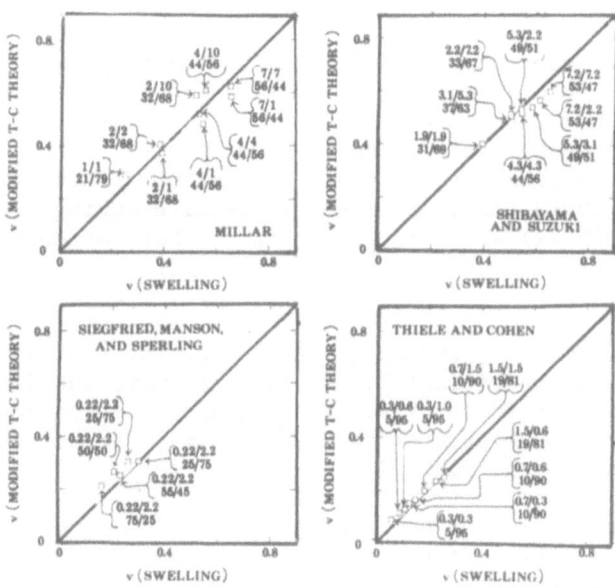

Fig. 3.   Swelling values predicted by the modified Thiele-Cohen
equation <u>vs</u> experimental swelling values.

substantial variations.  One way of viewing the data is that it
does not matter as much what the crosslink level of network II is
relative to network I.  The other series in the Millar (15) data
substantiate this finding, albeit less dramatically.  A statistical
analysis of the Shibayama-Suzuki (16) data indicates a similar trend.
Neither the data of Thiele and Cohen nor the data of Siegfried,
Manson and Sperling (14) support this conclusion.  The results of
this analysis are summarized in Table II.

Modulus Data

     A similar analysis was given to the modulus data, see Figures
4-6.  Figures 4a, 5a, and 6a show the predicted Young's modulus
based on stoichiometry, E(Stoichiometry) vs the experimental Young's
modulus, E(Experiment).  Figures 4b, 5b, and 6b show the modulus
predicted by equation (8) vs E(Experiment).  As with the swelling
data, the network imperfections and the contributions of the
physical crosslinks, if any, were minimized by determining the two
crosslink levels required for E(theory) on the separate homopolymer
networks.  Unfortunately, Millar did not report modulus data.

     Data lying to the right of the theoretical line provide an
indication of physical crosslinks, since the experimental modulus is
larger than the theoretic modulus.  A vertical stacking of points
containing the same network I (but different network II's), suggest
the dominance of network I over network II, since the experimental
modulus is relatively constant.

     The data, unfortunately, are observed to be erratic and
inconsistent with one another.  Shibayama and Suzuki's data, Fig. 4a,
show a reasonable fit of stoichiometry to experiment.  In Fig. 4b,
the preponderance of data to the right of the theoretical diagonal
suggest added physical crosslinks.  Domination of network I over
network II is suggested by a slight vertical stacking noted in the
3.1/3.1 (DVBI/DVBII) and 3.1/5.3 sequence.  More evident is the
horizontal stacking caused by inverting networks I and II, note the
3.1/5.3, 5.3/3.1 pair and the 2.2/7.2, 7.2/2.2 pair.

     Siegfried, Manson, and Sperling's data, Fig. 5a, all fall to
the right of the theoretical line.  Fig. 5b shows the data to the
left of the line, vertically stacked.  (All of the polymer network I
compositions have the same crosslink density.)  Hence network I is
assumed to dominate network II.

     In Figure 6a, Thiele and Cohen's data all lie to the left of
the theoretical line.  Perhaps the second vinyl in the divinyl
benzene was not entirely consumed.  The data in Fig. 6b fit the
theory surprisingly well, however, with a slight indication of
vertical stacking.

Table II. IPN Swelling Factors Unaccounted for by Theory

| Investigator | Added Physical Crosslinks? | Network I Domination? |
|---|---|---|
| Millar | no | yes |
| Shibayama and Suzuki | no | slight |
| Siegfried, Manson and Sperling | no | no |
| Thiele and Cohen | no | no |

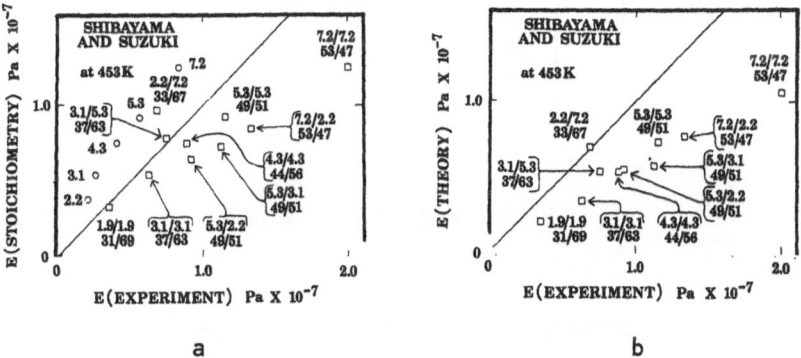

a                 b

Fig. 4. Young's modulus data of Shibayama and Suzuki. (a) The stoichiometric modulus assumes an average crosslinking level in space, computed from the DVB concentration contributed from each network (b) Young's modulus according to equation (8) vs experiment.

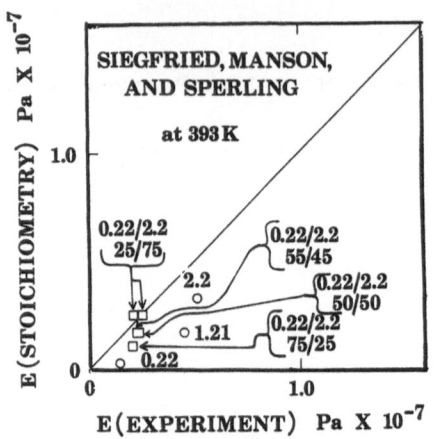

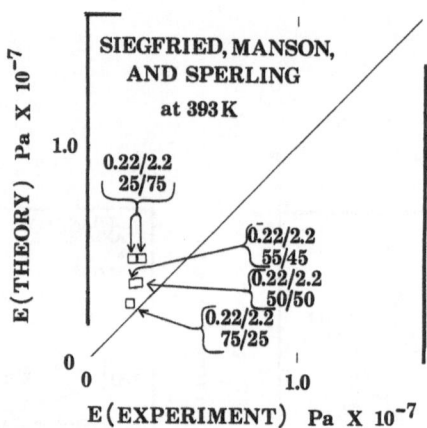

Fig. 5.  Young's modulus data of Siegfried, Manson, and Sperling.

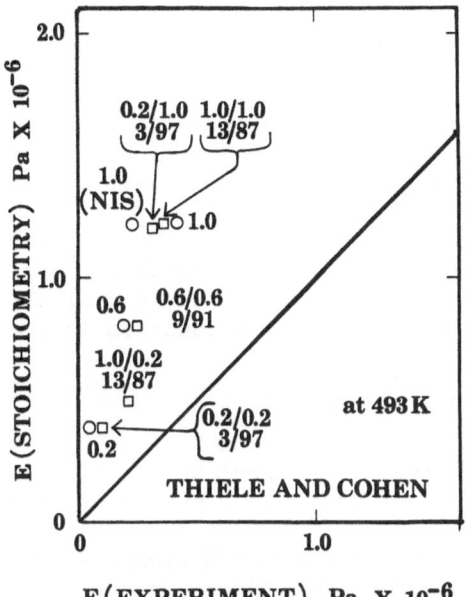

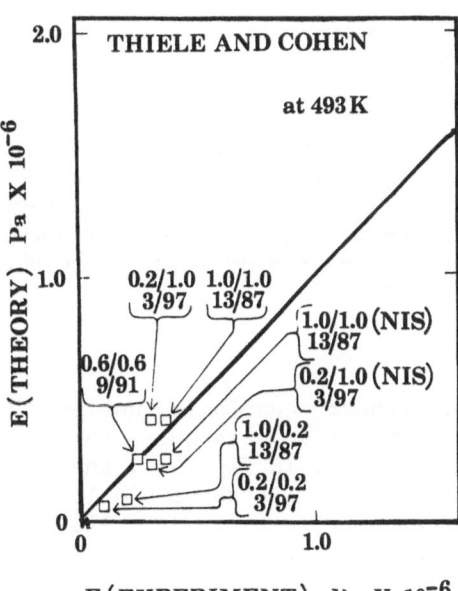

Fig. 6.  Young's modulus data of Thiele and Cohen.

The effects of network I domination and added physical crosslinks are summarized in Table III. Unfortunately, Millar presented no modulus data as his materials were suspension-sized particles.

Before any discussion of the results, two comments must be made. First, a much greater scatter of the data was encountered for the modulus data, compared to the swelling data. Second, in some cases, the values of the modulus were below those predicted, for single networks and IPN's alike. Because the data were collected over a range of temperatures (all above 100°C), it was more convenient to work in terms of the crosslink levels rather than in terms of the moduli. From equation (8), a value of

$$N(\text{theory}) = v_1^{1/3} N_1 + v_2 N_2 \qquad (12)$$

was defined. [In itself, N(theory) is an effective crosslink density and not representative of a measurable number of crosslinks.]

If the experimental values of N from E = 3NRT exceeded N(theory), this was taken as evidence for new physical crosslinks caused by IPN formation. Likewise, a slower than expected variation within a composition series or near series (via statistical analysis) was taken as evidence for an outsized contribution of network I.

While Table III affirms network I domination, there appears to be conflicting evidence for added physical crosslinks.

In addition to the swelling and modulus data analyzed above, the creep data and morphology (electron microscopy) studies by Siegfried, Manson and Sperling point to a greater continuity of network I. In the language of Lipatov and Sergeeva (12), network II appears to behave somewhat like a filler for network I. This is all the more surprising in the present case, where both polymers are identical in composition.

DISCUSSION

While the Flory-Rehner equation is not expected to fit homo-IPN data, it is of interest to note where the data lies with respect to the theoretical diagonal. In Figure 1, the data of Millar, and Shibayama and Suzuki tend to lie above the line, while Thiele and Cohen and Siegfried et al's. data tend to lie below it. Yet the Thiele-Cohen equation, applied to the same data in Fig. 2, showed a marked improvement, pulling the data towards the theoretical line whether it originally lay above or below the Flory-Rehner line in Fig. 1. The modified Thiele-Cohen equation, employed in Fig. 3, yields a somewhat better fit.  It remains for the future to ascertain whether the modification will be valuable with other data.

The fit for the modulus data was disappointing, with considerable scatter.  Also, the data were non-superimposable, one investigator to another.  Whether this was due to experimental error or to other factors remains to be seen.

It should be pointed out that both viscoelastic data and transmission electron microscopy data by Siegfried, et al. (14), not analyzed above, also support the speculation that network I dominates network II.  In fact, it is seen from the electron micrograph in Fig. 7 (14), that network I apparently exhibits greater continuity in space.  Network II is seen to form small domains about 75Å in diameter within network I.

Overall, network I domination is apparently a real phenomenon, but is much more obvious in the solid state than in the swollen state.  Perhaps the networks behave much more "ideally" when swollen which is not surprising.

The idea of network I dominating the properties of the IPN through its greater continuity in space has some important implications in thermoset resin synthesis, such as epoxy materials. The suggestion is that if a mixture of monomers are simultaneously polymerizing, then that portion of the material already incorporated in the network at the time of gelation may tend to be more continuous in space and dominate the properties.  That material polymerized later in time, statistically, may form less continuous domains and act like filler to a greater or lesser extent.  Partial confirmation of this concept has already been obtained in this laboratory (29,30).

While there is some evidence for new physical crosslinks, unfortunately the data do not permit a reasonable conclusion either way.  Only the modulus data of Shibayama and Suzuki indicates the presence of any added physical crosslinks contributing to the rubbery modulus.  Siegfried, Manson, and Sperling's data, Fig. 5b, indicate fewer  physical crosslinks in the homo-IPN than in the corresponding single networks, because the data lies to the left of the line.

Flory has concluded (31) that physical crosslinks should not contribute to the rubbery modulus.  Therefore, added physical crosslinks arising though interpenetration should also not contribute. So then the modulus should not reflect the number of the physical crosslinks.  However, other workers (2-5) find physical crosslinks making important contributions, if through various routes.  Clearly more work is required.

Table III.   IPN Modulus Factors Unaccounted for by Theory

| Investigator | Added Physical Crosslinks? | Network I Domination? |
|---|---|---|
| Millar | N.A. | N.A. |
| Shibayama and Suzuki | yes | yes |
| Siegfried, Manson and Sperling | no | yes |
| Thiele and Cohen | no | very slight |

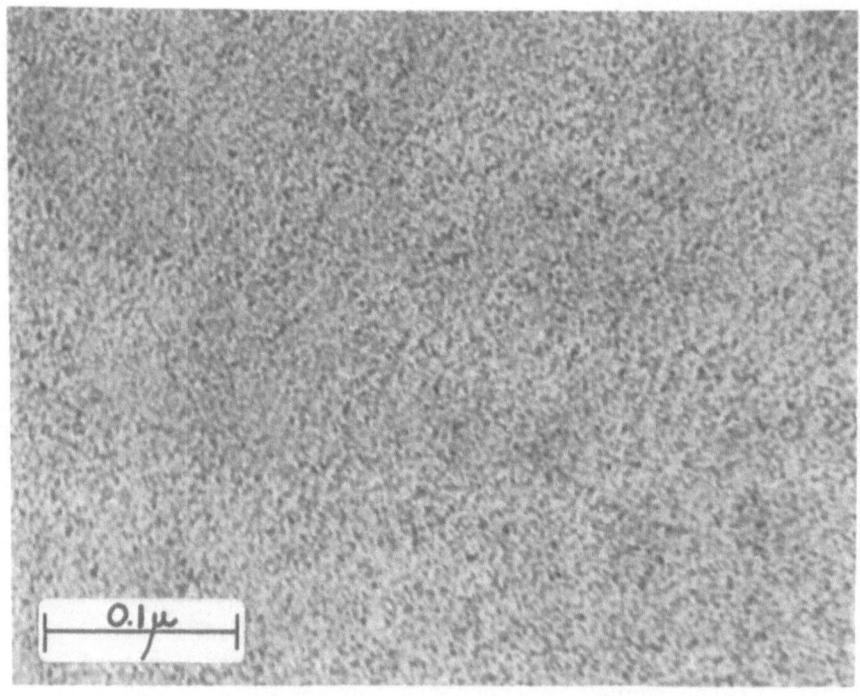

Fig. 7.   Morphology of 50/50-0.4/4% DVB + 1% isoprene homo-IPN. Polymer network II, darker regions stained with $OsO_4$, appear as domains near 75Å in diameter.

REFERENCES

1.  P. J. Flory, Proc. R. Soc. London, A., 351, 351-380 (1976).
2.  O. Kramer, V. Ty, and J. D. Ferry, Proc. Nat. Acad. Sci U. S.,
    69, 2216 (1972).
3.  R. L. Carpenter, O. Kramer, and J. D. Ferry, Macromolecules,
    10, 117 (1977).
4.  P. J. Flory, Trans. Faraday Soc., 56, 772 (1960).
5.  O. Kramer, in Europhysics Conference Abstracts, 3C, "Structure
    and Properties of Polymer Networks", European Physical Society,
    Warsaw, Poland, April 1979.
6.  W. W. Graessley, Adv. Polym. Sci., 16 (1974).
7.  H. M. James and E. Guth, J. Chem. Phys. 15, 669 (1947).
8.  B. N. Kolarz, J. Polym. Sci., 47, 197 (1974).
9.  W. Trochimczuk, in Europhysics Conference Abstracts, 3C,
    "Structure and Properties of Polymer Networks", European
    Physical Society, Warsaw, Poland, April 1979.
10. F. G. Hutchinson, R. G. C. Henbest, M. K. Leggett, U.S. 4, 062,
    826 (1977)
11. H. L. Frisch, K. C. Frisch, and D. Klempner, Mod. Plast., 54, 76
    (1977); ibid. 54, 84 (1977).
12. Yu. S. Lipatov and L. M. Sergeeva, Russian Chem. Rev., 45 (1),
    63 (1976).
13. D. A. Thomas and L. H. Sperling, Chapter 11 in Polymer Blends,
    D. Paul and S. Newman, Eds., Academic Press, 1978.
14. D. L. Siegfried, J. A. Manson and L. H. Sperling, J. Polym. Sci.,
    Polym. Phys. Ed., 16, 583 (1978).
15. J. R. Millar, J. Chem. Soc., 1311 (1960).
16. (a) K. Shibayama and Y. Suzuki, Kobunshi Kagaku, 23, 24 (1966);
    (b) Rubber Chem. Technol., 40,476 (1967).
17. J. L. Thiele and R. E. Cohen, Polym. Eng. Sci., 19, 284 (1979).
18. H. A. Clark, U.S. Pat. 3,527,842 (1970).
19. J. J. P. Staudinger and H. M. Hutchinson, U.S. Pat. 2,539,377
    (1951).
20. Yu. S. Lipatov, V. F. Rosovizky and V. F. Babich, European
    Polym. J., 13, 651 (1977).
21. K. J. Smith and R. J. Gaylord, J. Polym. Sci., A-2, 10, 283
    (1972).
22. S. C. Kim, D. Klempner, K. C. Frisch, W. Radigan, and H. L.
    Frisch, Macromolecules, 9, 258 (1976).
23. N. Devia-Manjarres, J. A. Manson, L. H. Sperling, and A. Conde,
    Polym. Eng. Sci., 18, 200 (1978).
24. A. V. Tobolsky and M. C. Shen, J. Appl. Phys., 37, 1952 (1966).
25. M. C. Shen, T. Y. Chen, E. H. Cirlin, and H. M. Gebhard, in
    "Polymer Networks: Structure and Mechanical Properties," A. J.
    Chompff and S. Newman, Eds., Plenum, New York, 1971.
26. B. Meissner and I. Klier, in Europhysics Conference Abstracts,
    3C, "Structure and Properties of Polymer Networks", European
    Physical Society, Warsaw, Poland, April 1979.

27.  P. J. Flory and J. Rehner, J. Chem. Phys., 11, 512 (1943).

28.  A. V. Galanti and L. H. Sperling, Polym. Eng. Sci., 10, 177 (1970).

29.  S. C. Misra, J. A. Manson, and L. H. Sperling, ACS Org. Coat. and Plast. Chem. Preprints, 39(2), 146 (1978).

30.  S. C. Misra, J. A. Manson, and L. H. Sperling, ACS Org. Coat. and Plast. Chem. Preprints, 39(2), 152 (1978).

31.  P. J. Flory, in Europhysics Conference Abstracts, 3C, "Structure and Properties of Polymer Networks", European Physical Society, Warsaw, Poland, April 1979.

# POLYURETHANE-ACRYLIC COPOLYMER PSEUDO INTERPENETRATING

# POLYMER NETWORKS

D. Klempner, H. K. Yoon, K. C. Frisch
Polymer Institute, University of Detroit, Detroit, MI 48221

H. L. Frisch
Dept. of Chemistry, State University of New York
Albany, NY

## INTRODUCTION

Previous investigations (1-6) have demonstrated that inter-penetrating polymer networks (IPN's) have exhibited better mechanical properties than those of their component networks.  In addition, more complete phase mixing was observed for IPN's than for blends of the corresponding linear polymers.  These results were interpreted in terms of the permanent interlocking of the polymer chains of the composite structure.  In this study, three different types of polymer blends, i.e., linear blends, pseudo-IPN's and IPN's, were prepared from a polyurethane and an acrylic copolymer.  The polymers studied were similar to those used in an earlier study (7,8).  However, the polyacrylate used in the present study cross-links by means of a free radical mechanism, while the polymer used previously was crosslinked with a melamine-formaldehyde resin via pendant hydroxyls.  Therefore, in this study the possibility of grafting occurring between the component polymers is much less (i.e., reaction of the melamine with the hydroxyl-terminated chain extender for the polyurethane).  The pseudo-IPN's, made from a linear polymer and a crosslinked polymer, would theoretically result in temporary entanglements rather than permanent entangle-ments, which occur in the IPN's.

The mechanical properties and glass transition behavior of these polymer alloys were compared.  The kinetics of polymeri-zation of the component polymers were measured and varied by changing the concentration of catalysts in order to determine the effect of polymerization rates on the morphology of the IPN's. Electron microscopy and dynamic mechanical spectroscopy were also carried out.  Several theoretical models predicting the modulus of

the heterogeneous systems were tried in order to fit the experimental
results.   In addition, the density behavior of these materials was
determined and interpreted in terms of the degree of phase mixing
(interpenetration).

EXPERIMENTAL

Materials

Poly(1,4-oxybutylene)glycol, M.W. = 1017, (Polymeg 1000) was
dried at $80^{\circ}C$ for five hours under a vacuum of 1 mm Hg.  All vinyl
monomers were purified by washing with an 0.5% sodium hydroxide
solution  and vacuum distilled at 1 mm Hg.

Preparation of Networks

Polyurethane (PU) Prepolymer:  The prepolymer (NCO/OH = 2:1)
was prepared at $80^{\circ}C$ under nitrogen.   508.6 g (1 equivalent) Polymeg
1000 (M.W. = 1017.2) were added slowly with stirring to 262 g (2
equivalents) of 4,4'-methylene bis(cyclohexy isocyanate) (Hylene
W) in a one liter resin kettle.  The reaction was carried out until
the theoretical isocyanate content (determined by the di-n-buty
amine method) (9) was reached.

Acrylate Copolymer (AC):  A copolymer of eight parts of n-butyl
methacrylate, one part of ethyl methacrylate and one part of styrene
was prepared.   800 g of n-butyl methacrylate, 100 g of ethyl methacrylate and 100 g of styrene were heated with stirring in a 2000
cc three-necked flask, equipped with a reflux condenser and nitrogen inlet.   The reaction was carried out at $80^{\circ}C$ until a prepolymer
of syrupy consistency was obtained.  The polymer was very similar
in nature to the polyacrylate used in a previous study (7,8),
except that it contained no pendant hydroxyl or carboxyl groups
needed for the melamine-formaldehyde cure.

Linear Blends:  The PU prepolymer and 1,4-butanediol were
mixed thoroughly with the acrylate copolymer syrup.  0.2% T-12
(by wt. of PU) and 1% benzoyl peroxide (by wt. of AC) were used as
catalysts.  The mixture was cast between two glass plates provided
with rubber seals around the edges, and subsequently polymerized
at $110^{\circ}C$ for 24 hours.  Combinations composed of 20%, 40%, 60% and
80% PU by weight were made.

Pseudo-IPN's (PDIPN's):  Two types of pseudo-IPN's were prepared, one from a linear PU/crosslinked AC (PDIPN-2), and the
other from a linear AC/crosslinked PU (PDIPN-1).  The former
pseudo-IPN's were prepared from the mixture of the linear polyurethane component (PU prepolymer + 1,4-butanediol) and the acrylate copolymer syrup with ethylene glycol dimethacrylate (EGDMA).
The molecular weight between crosslink sites (Mc) was 6500.  The

mixture of acrylate copolymer syrup and the PU prepolymer admixed
with 1,4-butanediol and trimethylolpropane (crosslinking agent)
resulted in the latter pseudo-IPN. The curing conditions and compositions were the same as in the linear blends.

IPN's: Full IPN's (FIPN's) were prepared from the mixture of
acrylate copolymer syrup with EGDMA and PU prepolymer with 1,4-
butanediol and trimethylolpropane. The curing conditions and compositions were the same as those for the linear blends. The crosslink densities of both polymers were the same (Mc = 6500).

## Measurements

Stress-Strain Properties: The tensile strengths and elongations at break were measured on an Instron Tensile Tester at room
temperature with a crosshead speed of 2 in/min. Specimens were
0. 125 inch wide dumbbells.

Glass Transition Temperature: The glass transition temperatures were measured on a Perkin-Elmer Differential Scanning Calorimeter, DSC-2. Measurements were carried out from $-120^{\circ}C$ to $+60^{\circ}C$
under helium at a scanning rate of $10^{\circ}C$ per min. Specimen sizes
were on the order of 20 mg.

Hardness: The hardness of IPN's was measured by means of
Shore A and D durometers according to ASTM D 2240-75.

Kinetic Measurements: Three different catalyst concentrations
were employed for each polymer, i.e., 0.5%, 1% and 2% of benzoyl
peroxide in the acrylate copolymer; 0.02%, 0.1% and 0.5% of T-12 in
the polyurethane. One g of each sample was poured into a glass
tube (12 cm long, 8 mm diameter). The tubes were subsequently
sealed. These sample capsules were heated in an oil bath at $110^{\circ}C$.
The samples were taken out at intervals and were quenched in a dry
ice-acetone bath and examined for flowability. The gel time was
taken as the time at which the polymerizing mass did not flow at
room temperature.

Electron Microscopy: The sample preparation was based on
Kato's (10) osmium tetroxide staining technique and a two-step
sectioning method. The specimens were exposed to $O_5O_4$ vapor and
cut with a LKB ultratome III to get a 0.1 $\mu$ slice. The electron
micrographs were taken with an AEI 6B and a Phillips 300 transmitting electron microscope with a magnification of 95,000.

Mechanical Spectroscopy: Dynamic viscoelastic properties were
obtained using a Rheovibron Model DDV-11 (manufactured by Toyo
Measuring Instruments Co., Ltd., Tokyo, Japan). The measurements
were taken over a temperature range of $-90^{\circ}C$ to $80^{\circ}C$ using a frequency of 110 Hz, and a heating rate of $2^{\circ}C/min$. Sample dimensions

were 0.024 x 0.6 x 3.5 cm.

Density:  The density was measured at room temperature using
a hydrostatic technique.  The sample dimensions were 0.125 x 1.25
x 2.5 cm.

RESULTS AND DISCUSSION

Morphology

The morphology, as indicated by the glass transition behavior
(Table I) was heterophase.  In all cases, two $T_g$'s, corresponding
to the $T_g$'s of the component polymers, resulted.  However, the $T_g$'s
were shifted inwards, indicating some phase mixing (interpenetration), most likely at the phase boundaries.  The shifts were greatest with the FIPN's, indicating the greatest amount of interpenetration.  In the case of the full IPN's, intermediate $T_g$'s occurred
(not shown in the table) depending on the thermal history of the
material.  This would indicate a third phase, i.e., an interpenetrating region.  The PDIPN's and linear blends showed no indication
of this intermediate phase.  Further studies should clarify this
situation.

---

TABLE I

GLASS TRANSITION TEMPERATURES (°C)

|              | PU 100% PA 0% | 80 20 | 60 40 | 40 60 | 20 80 | 0 100 |
|--------------|---------------|-------|-------|-------|-------|-------|
| FIPN         | 204           | 214   | 211   | 209   | 210   | 320.5 |
|              | –             | 306   | 306   | 308   | 312   |       |
| PDIPN-1      | 204           | 211   | 212   | 212   | 210   | 318   |
|              | –             | 313   | 317   | –     | –     |       |
| PDIPN-2      | 204           | 211   | 212   | 213   | 212   | 320.5 |
|              | –             | 313   | 314   | –     | –     |       |
| Linear Blend | 204           | 212   | 208   | 205   | 208   | 318   |
|              | –             | 320   | 319   | –     | –     |       |

The electron micrographs of the IPN's, PDIPN's and linear
blends (Figures 1-4) all showed heterogeneous behavior, in agreement with the $T_g$ results.  This morphology was expected due to the
differing solubility parameters of the two polymers.  Considering
the inwardly shifted glass transition temperatures obtained by DSC
(Table I), some phase mixing (interpenetration) resulted in all
cases.  It can be seen that the phase boundaries of the IPN's and
PDIPN's-2 were not well defined while the PDIPN's-1 and linear blends
showed well defined boundaries.  In all cases the blends having a
crosslinked acrylic component exhibited more interpenetration.  A
possible explanation would be the ease of diffusion of the growing
chains.  The diffusion rates of the linear and crosslinked polyurethane do not change rapidly during polymerization because the
initial prepolymer viscosity is very high.  However, the viscosity
of the crosslinked acrylic copolymer increases faster than that of
the linear acrylate copolymer.  Thus the crosslinked acrylic copolymer does not have enough diffusion time to allow complete phase
separation.  This is consistent with the glass transition behavior
being dependent upon the polymerization rates discussed below.
Phase inversion was observed to occur around 60% polyurethane in
all cases.  The domain sizes in IPN's and PDIPN's-2 were about the
same and ranged from 100A to 500A.  However, the PDIPN-1 and linear
blends showed larger domain sizes (500A - 5000A) indicating more
complete phase separation (i.e., less interpenetration).  This demonstrates the importance of topology and kinetics on the morphology
and degree of interpenetration of these alloys.

Dynamic Mechanical Properties

The dynamic storage modulus, E', vs. temperature for the
IPN's, PDIPN's and linear blends show that the general modulus
response to acrylic concentration and temperature for all the
alloys are alike (figures not shown).  The modulus of the IPN's,
the PDIPN's-1, the PDIPN's-2, and also for the linear blends decreases systematically with increasing polyurethane concentration.
This can be seen in Figures 5 and 6, which show the modulus vs.
acrylic composition for a full and pseudo-IPN.  The largest jump
of modulus takes place between 60% and 40% PU, indicating the phase
inversion.  The modulus of the alloys increased in the order of
linear blends <. PDIPN's-2 < PDIPN's-1 < FIPN's at corresponding
concentrations.  The high degree of phase mixing did not result in
an increase of the modulus of these blends; electron micrographs
showed more phase mixing in the case of PDIPN's-2 than PDIPN's-1.
The damping factor, tan $\delta$ vs. temperature for the pseudo and full
IPN's showed two peaks corresponding to the $T_g$'s of the polyurethane and the acrylic copolymer networks.  The size of the damping
peaks changed systematically with concentration of the alloys,
i.e., decreased with the component concentration.

The modulus-composition behavior of the polyurethane acrylic

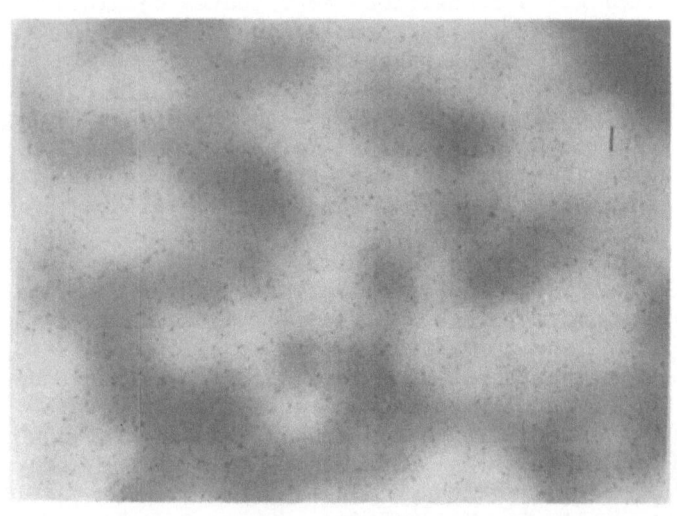

Figure 2.    Electronmicrograph
of UL40AC60 (PDIPN-2)

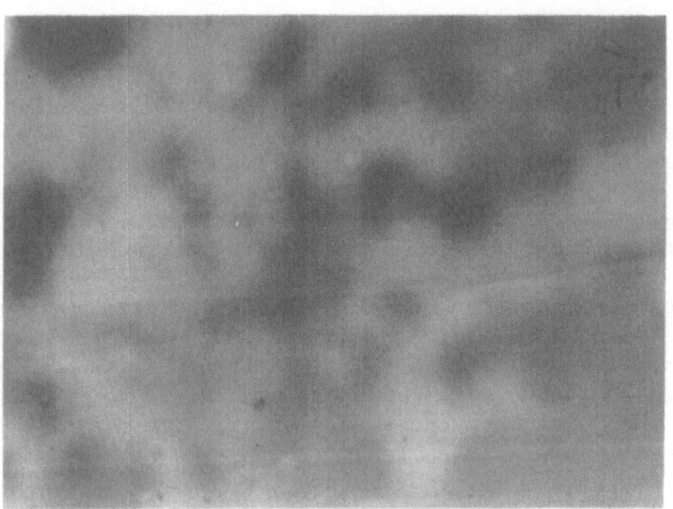

Figure 1.    Electronmicrograph
of UC20AC80 (IPN)

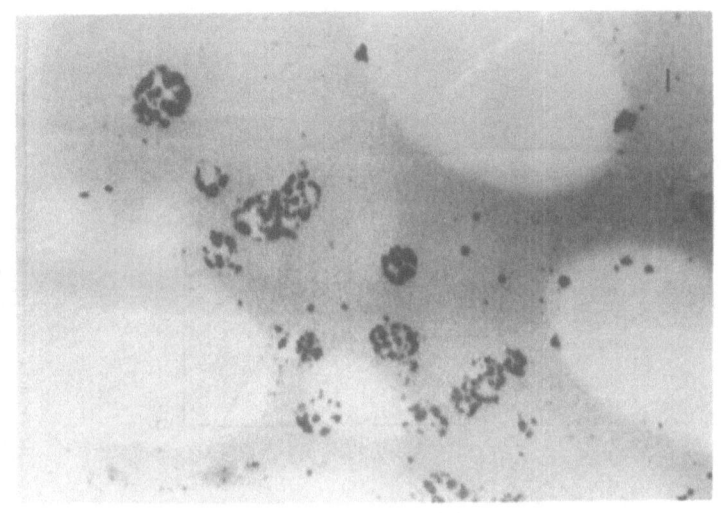

Figure 4. Electronmicrograph of UL60AL40 (linear blend)

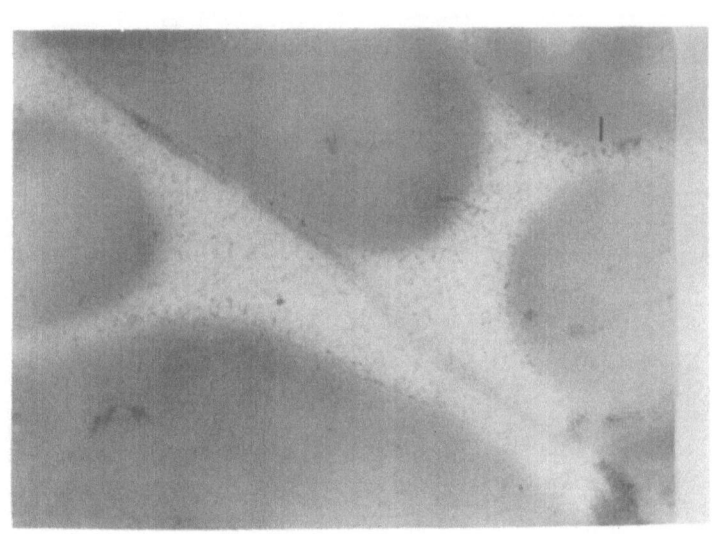

Figure 3. Electronmicrograph of UC40AL60 (PDIPN-1)

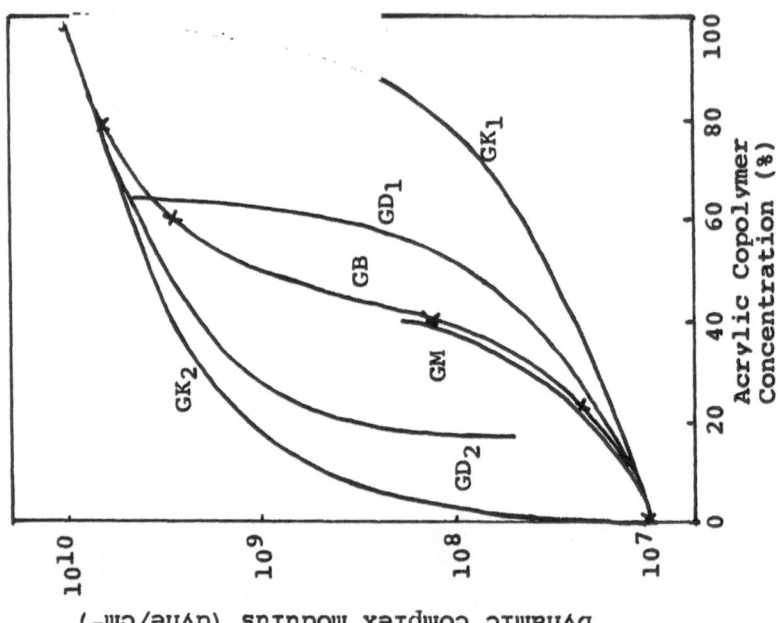

Figure 6.

Dynamic Complex Modulus vs. Acrylic Copolymer Concentration for FIPN (X) at 23° at 110 Hz Frequency (Solid lines are based on theoretical models).

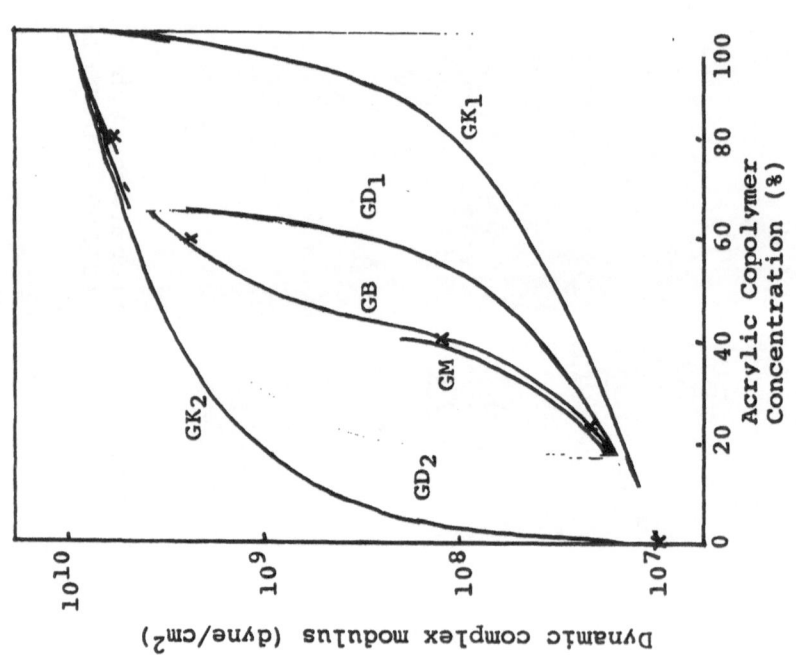

Figure 5.

Dynamic Complex Modulus vs. Acrylic Copolymer Concentration for PDIPN (X) at 23° C at 110 Hz Frequency (Solid lines are based on theoretical models).

copolymer IPN's was analyzed with some of the theoretical equations based on mechanical models. Most of these models assumed perfect adhesion between the phases and a spherical inclusion (filler) geometry. They were discussed and reviewed by Nielsen (11), Dickie (12-15) and Davies (16).

Mooney (17) developed relations based on Einstein's theory (18) for elastomers having a Poisson's ratio of 0.5 and filled with rigid spherical fillers.

$$G = G_m \exp \frac{2.5\ V_i}{1 - V_i/\emptyset_m} \tag{1}$$

where $G_m$ is the shear modulus of the elastomers (the matrix), $V_i$ is the volume fraction of the filler (the inclusion), and $\emptyset_m$ is the maximum volume fraction that the filler can have due to packing difficulties from particle-particle contact.

Substituting tensile modulus, E, for G in the Mooney equation by using the relation $E = 2\ (1 + \nu)G$, it follows that:

$$E = \frac{(1 + \nu)}{(1 + \nu_m)}\ E_m \exp \frac{2.5\ V_i}{1 - V_i/\emptyset_m} \tag{2}$$

where $\nu$ and $\nu_m$ are the Poisson's ratios for the composite and for the elastomer (the matrix), respectively.

Kerner (19) developed relations for the gross bulk and shear modulus of a multicomponent system. For two components his equation may be written as:

$$\frac{G}{G_m} = \frac{(1 - V_i)G_m + (\alpha + V_i)G_i}{(1 + \alpha V_i)G_m + \alpha(1 - V_i)G_i} \tag{3}$$

where G is the shear modulus, $V_i$ is the volume fraction of the inclusions (fillers), subscript $m$ denotes a matrix property, subscript $i$ denotes an inclusion property, and $\alpha$ is a function of the Poisson's ratio of the matrix:

$$\alpha = 2(4 - 5\nu_m)/(7 - 5\nu_m) \tag{4}$$

In terms of tensile modulus E:

$$\frac{E}{E_m} = \frac{\gamma(1 - V_i)E_m + \beta(\alpha + V_i)E_i}{(1 + \alpha V_i)E_m + \alpha\beta(1 - V_i)E_i} \tag{5}$$

where

$$\beta = (1 + \gamma_m)/(1 + \gamma_i) \tag{6}$$

$$\gamma = (1 + \gamma_i)/(1 + \gamma_m) \tag{7}$$

Equivalent equations were also developed by Takayanagi (20).

Dickie (12) modified eq 5 and introduced an interaction parameter $\psi$ in the form of a maximum packing fraction $\emptyset_m$:

$$\frac{E}{E_m} = \frac{\gamma(1 - \psi V_i)E_m + \beta(\alpha + \psi V_i)E_i}{(1 + \alpha\psi V_i)E_m + \alpha\beta(1 - \psi V_i)E_i} \tag{8}$$

where

$$\psi = 1 + V_i(1 - \emptyset_m)/\emptyset_m^2 \tag{9}$$

The above theories are based on a polymer composite where the matrix and the dispersed phase are well defined, i.e., the matrix and the dispersed phase are both single components of the system. The properties of the matrix material are dominant in determining the composite properties. Theoretically, the equations will represent the composite properties better if the volume fraction of the dispersed phase is small, since they implicitly assume no interaction between the dispersed phase domains.

The distinction between the matrix and the inclusion disappears in the range of intermediate concentrations where the phase inversion occurs. There is also large interaction between the phase domains and one component network does not play a dominant role over the other in determining the composite properties. At these concentrations, the previous theories only give upper and lower bounds of the composite modulus.

In order to account for the interaction between the phase domains, Budiansky (21) used the Kroner's "smearing out" approach (22) and assumed that an isolated spherical inclusion is embedded in an infinite isotropic matrix which itself is a composite material. For two-phase systems, his equation is expressed as:

$$\frac{V_1}{1 + \varepsilon(G_i/G - 1)} + \frac{V_2}{1 + \varepsilon(G_2/G - 1)} = 1 \tag{10}$$

where

$$\varepsilon = \frac{2(4 - 5\gamma)}{15(1 - \gamma)} \tag{11}$$

Subscripts 1 and 2 denote component materials and $\gamma$ is the
Poisson's ratio of the composite. It is interesting to note that
eq 10 is symmetrical and the reversal of geometrical roles of the
materials 1 and 2 does not change the composite property in a
given concentration. In other words, one does not differentiate
materials 1 and 2 as the matrix or the inclusion. The Mooney
equation (eq 1 and 2), the Kerner equation (eq 3 and 5), the Dickie
equation (eq 8) ($\gamma = 1$ for shear modulus, G, in eq 8) and the
Budiansky equation (eq 10) were compared with the experimental
results at 23°C. The Poisson's ratio of the polyurethane phase
was assumed to be 0.5 and that of the acrylic to be 0.35 at 23°C.
The Poisson's ratio of the composite was assumed to be the volume
additive of its components:

$$\gamma = v_m \gamma_m + v_i \gamma_i$$

Although this assumption is not theoretically valid, the resulting
ratios,

$$\frac{1 + \gamma}{1 + \gamma_m} \text{ and } \frac{1 + \gamma}{1 + \gamma_i}$$

in eq 2, 5 and 8, varied at the most $\pm$ 5%. The maximum packing
fraction, $\emptyset_m = 0.64$, was chosen for rigid inclusions in an elastomeric matrix, while $\emptyset_m = 0.83$ was used for elastomeric inclusions
in a rigid matrix (14,16).

   The modulus-composition plots and their comparison with the
models are shown in Figs. 5 and 6. Each model predicts the modulus
fairly well within a narrow range of concentration. However, over
the entire concentration range the experimental values fit the
Budiansky model (21) best, which was expected, since the assumptions made in the model represent the phase inversion process.
This agrees well with a previous finding with similar polymer networks (23). An interesting result is that the experimental values
for the FIPN's are slightly higher than the Budiansky values,
while the other blends show lower values than the model. This indicates the reinforcement effect achieved only with full IPN's.

## Density

   The density-composition curve (Fig. 7) shows increased density
in all cases, maxima occurring at 80% PU. The IPN's exhibited the
largest increase at all concentrations. This increase seems to
indicate increased molecular mixing in full IPN's. This can be
verified by the glass transition temperature behavior. Kim et al
explained the increased density of IPN's qualitatively by means of
chain entanglements at the domain boundaries (24). However, it is
possible to quantitatively correlate it with $T_g$ shift. As the
glass transition temperatures of the blends shift inward (from

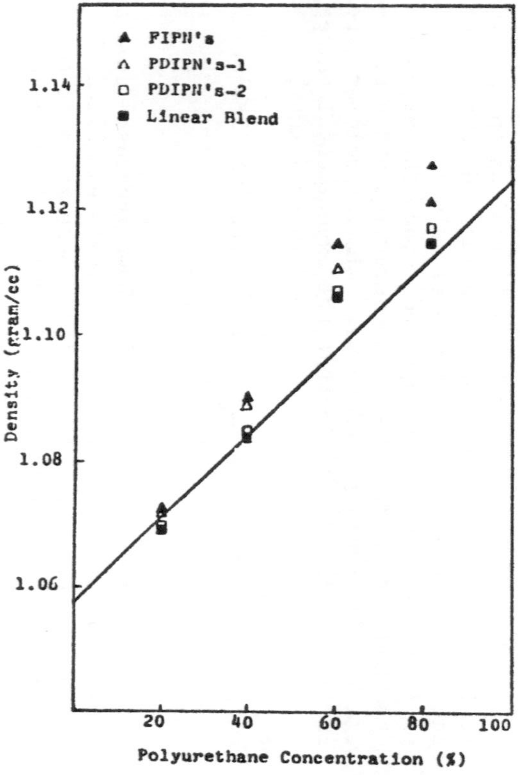

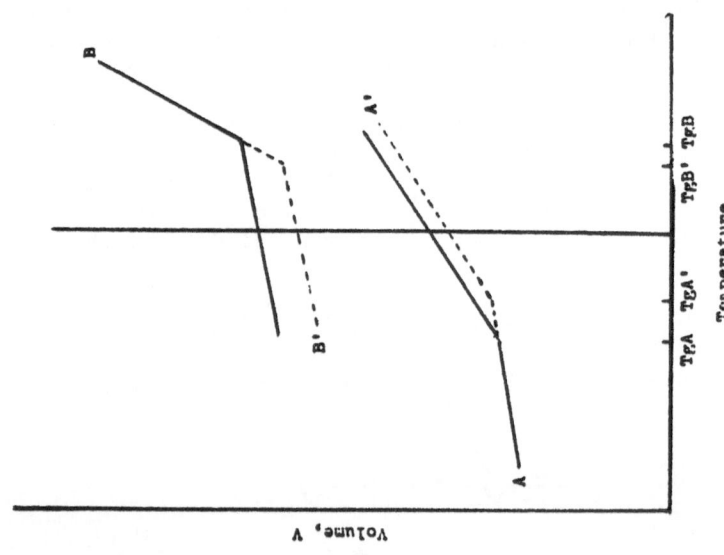

Figure 8.   Volume vs. Temperature Curves of Pure Homopolymers and Polymer Blends.

Figure 7.   Density vs. Polyurethane Concentration for PU-PA (Straight line is based on volume additivity).

$T_gA$ and $T_gB$ to $T_gA'$ and $T_gB'$, respectively), the volume-temperature
curves (Fig. 8) of the components in the blends follow the dotted
lines instead of the solid lines. Thus, the volume of each component becomes smaller than that of the pure component, which results in higher density at room temperature. Note that the densification of the IPN's is due to interpenetration, which produced
the $T_g$ shift and that the effect was the greatest with full IPN's,
in agreement with glass transition behavior and electron microscopy. Thus, it may be possible to predict the glass transition
behavior of the polymer blends by measuring the densities at several temperatures.

## Mechanical Properties

The tensile strengths and breaking elongations of the IPN's
as a function of composition are shown in Tables II and III, respectively. The IPN's show a significant enhancement in tensile
strength, particularly at PU/AC = 80:20. The PDIPN's and the linear blends show effectively no enhancement. In all cases measured,
the tensile strengths of the IPN's are greater than those of the
pseudo-IPN's and the linear blends at the same compositions. In
addition, the tensile strengths of both of the pseudo-IPN's are
greater than those of the linear blends at the same compositions.
The enhancement in tensile strength is most likely due to the
higher level of mixing in the IPN's and to greater adhesion between the dispersed and continuous phases. The IPN's should exhibit more complete mixing (less phase separation) and greater
phase adhesion due to the permanent entanglements (interpenetration)
at the phase boundaries. However, in the case of pseudo-IPN's, the
phase mixing is less due to the presence of temporary entanglements, since only one polymer is crosslinked. In the case of the
linear blends, even less mixing would be expected, since there are
no topological constraints at all, for neither polymer is crosslinked.

The elongation, as shown in Table III, shows a slight increase
from 100% PU to 80% PU in the IPN's , indicating the exceptional
toughness of this IPN. No increase occurs for the PDIPN's and
the blends. A sharp drop in elongation takes place at 60% PU,
which indicates that a phase inversion occurs around this composition. The Shore A and D hardnesses shown in Table IV show a rapid
increase between 80% PU and 40% PU, which again indicates the phase
inversion between 80% and 40% PU content.

## Kinetic Measurements

The gel times of the polymers, shown in Table V, show the
expected trend, i.e., increasing catalyst concentration decreases
the gel time. The glass transition behavior of these IPN's
(Table VI) shows that as the rate of the respective polymerizations

TABLE II

TENSILE STRENGTH OF IPN'S (PSI)

| | PU 100%<br>AC 0% | 80<br>20 | 60<br>40 | 40<br>60 | 20<br>80 | 0<br>100 |
|---|---|---|---|---|---|---|
| Full IPN | 6100 | 7100 | 6525 | 5000 | 3715 | 2571 |
| PDIPN-1 | 6100 | 5875 | 6300 | 3127 | 2227 | 1613 |
| PDIPN-2 | 5600 | 4428 | 5215 | 3602 | 3018 | 2571 |
| Linear Blend | 5600 | 1828 | 3617 | 3413 | 2118 | 1610 |

TABLE III

ELONGATION OF IPN'S (% AT BREAK)

| | PU 100%<br>AC 0% | 80<br>20 | 60<br>40 | 40<br>60 | 20<br>80 | 0<br>100 |
|---|---|---|---|---|---|---|
| Full IPN | 640 | 780 | 540 | 270 | 180 | 15 |
| PDIPN-1 | 640 | 540 | 250 | 138 | 120 | 12 |
| PDIPN-2 | 610 | 553 | 520 | 276 | 211 | 15 |
| Linear Blend | 610 | 640 | 357 | 130 | 195 | 12 |

TABLE IV

HARDNESS OF IPN'S (SHORE A)

| | PU 100%<br>AC 0% | 80<br>20 | 60<br>40 | 40<br>60 | 20<br>80 | 0<br>100 |
|---|---|---|---|---|---|---|
| Full IPN | 68(32)* | 70(35) | 85(38) | 92(53) | 94(63) | 100(72) |
| PDIPN-1 | 68(32) | 78(33) | 86(38) | 94(49) | 95(55) | 100(68) |
| PDIPN-2 | 73(31) | 81(37) | 91(43) | 93(53) | 96(65) | 100(72) |
| Linear Blend | 73(31) | 75(30) | 87(35) | 92(48) | 95(58) | 100(68) |

*Hardnesses in parentheses are measured by Shore D.

increases, the $T_g$'s shift inward, indicating less phase separation.
This is most likely due to the fact that increasing the reaction
rate lowers the diffusion time allowed for phase separation.  The
initial prepolymer mixture is single phase, since the molecular
weights are low enough to allow complete mixing.  As the polymers
are cured, they phase separate due to their thermodynamic incompatibility.  This phase separation is naturally diffusion controlled and requires time.  If the curing reactions are accelerated,
less time will be available for diffusion and subsequent phase
separation to occur.  When fully crosslinked, further phase separation cannot take place due to topological constraints imposed by
the crosslinking of the component polymers.  Thus, faster crosslinking would be expected to yield less phase separation.  The
maximum shift in $T_g$ occurs in the IPN 3 series, in particular,
3-3.  In this case, both polymers gel at almost the same time.
This is in agreement with Sperling's results (25), which showed
that the smallest domain size of epoxy-acrylic IPN's occurred
when the respective curing reactions were closest to simultaneity.
The degree of phase mixing, as deduced by the shifts in $T_g$, are
confirmed by the transparency of the IPN films.  The transparency
increases as the reaction rate increases.

TABLE V

GEL TIMES OF CROSSLINKED POLYMERS AT
DIFFERENT CATALYST CONCENTRATIONS (min.)

| AC | Time | PU | Time |
|----|------|-----|------|
| AC-1 | 4.25 | PU-1 | 12.30 |
| AC-2 | 4.00 | PU-2 | 7.25 |
| AC-3 | 3.20 | PU-3 | 3.15 |

| | | | |
|---|---|---|---|
| AC-1 | (0.5% BPO) | PU-1 | (0.02% T-12) |
| AC-2 | (1% BPO) | PU-2 | (0.1% T-12) |
| AC-3 | (2% BPO) | PU-3 | (0.5% T-12) |

TABLE VI

GLASS TRANSITION TEMPERATURES OF IPN'S:
KINETIC RESULTS

| | AC 40 - PU 60 $T_g$ (°K) | | AC 60 - PU 40 $T_g$ (°K) | |
|---|---|---|---|---|
| IPN 1.1* | 212.2 | 318.5 | 212 | 319 |
| IPN 1.2 | 211 | 318 | 210 | 318 |
| IPN 1.3 | 215 | 314 | 214 | 314 |
| IPN 2.1 | 212 | 317 | 211 | 318 |
| IPN 2.2 | 212.5 | 316 | 214 | 315.5 |
| IPN 2.3 | 216 | 313 | 215 | 315 |
| IPN 3.1 | 211 | 314 | 211.5 | 316 |
| IPN 3.2 | 214 | 308 | 215 | 307 |
| IPN 3.3 | 219 | 293 | 220 | 298 |

*IPN 1.1 (IPN from AC-1 and PU-1).

REFERENCES

1.  H. L. Frisch, D. Klempner and K. C. Frisch, J. Polymer
    Sci. (A-2) 8, 921 (1970).
2.  M. Matsuo, T. K. Kwei, D. Klempner and H. L. Frisch, Polymer
    Eng. and Sci. 10, (6), 327 (1970).
3.  D. Klempner and H. L. Frisch, J. Polymer Sci. (B), 8, 525
    (1970).
4.  L. H. Sperling and D. W. Friedman, J. Polymer Sci. (A-2), 7,
    425 (1969).
5.  A. J. Curtius, M. J. Covitch, D. A. Thomas and L. H. Sperling,
    Polymer Eng. and Sci., 12, 101 (1972).
6.  V. Huelck, D. A. Thomas and L. H. Sperling, Macromolecules,
    5, 340 (1972).
7.  K. C. Frisch, D. Klempner, S. Migdal and H. L. Frisch, J.
    Polymer Sci. (A-1), 12 (4), 885 (1974).
8.  K. C. Frisch, D. Klempner, S. Migdal and H. L. Frisch, J.
    Appl. Polymer Sci., 19, 1893 (1975).

9.  E. J. Malec and D. J. David, "Analytical Chemistry of Poly-
    urethanes," D. J. David and H. B. Staley, Eds., Wiley,
    New York, 1969, 1087.
10. K. Kato, Japan Plastics, 2, 6 (1968).
11. L. E. Nielsen, "Mechanical Properties of Polymers and Com-
    posites," Vol. 2, Marcel Dekker, New York, NY, 1974,
    p. 365.
12. R. A. Dickie, J. Appl. Polym. Sci. 17, 45 (1973).
13. R. A. Dickie, M. F. Cheung and S. Newman, J. Appl. Polym.
    Sci., 17, 65 (1973).
14. R. A. Dickie and M. F. Cheung, J. Appl. Polym. Sci., 17,
    79 (1973).
15. R. A. Dickie, J. Appl. Polym. Sci., 17, 2509 (1973).
16. G. Allen, M. J. Bowden, S. M. Todd, D. J. Blundel, G. M.
    Jeffs and W. E. A. Davies, Polymer, 15, 28 (1974).
17. M. Mooney, J. Colloid Sci., 6, 162 (1951).
18. A. Einstein, Ann. Phys. (Leipzig), 19, 549 (1905); 19, 289
    (1906); 34, 591.
19. E. H. Kerner, Proc. Phys. Soc. London, Sect. B., 69, 808
    (1956).
20. M. Takayanagi, H. Harima and V. Iwata, Mem. Fac. Eng.,
    Kyushu Univ., 23, 1 (1963).
21. B. Budiansky, J. Mech. Phys. Solids 13, 223 (1965).
22. E. Kroner, Z. Phys., 151, 504 (1958).
23. S. C. Kim, D. Klempner, K. C. Frisch and H. L. Frisch,
    Macromolecules 10, 1, 1187 (1977).
24. S. C. Kim, D. Klempner, K. C. Frisch and H. L. Frisch,
    Macromolecules, 9, 263 (1976).
25. R. L. Touhsaent, D. A. Thomas and L. H. Sperling, J. Poly.
    Sci., 46, 175 (1974).

POLY(2,6-DIMETHYL-1,4-PHENYLENE OXIDE) POLYSTYRENE

INTERPENETRATING POLYMER NETWORKS

H. L. Frisch
Dept. of Chemistry, State University of New York
Albany, NY

D. Klempner, H. K. Yoon, K. C. Frisch
Polymer Institute, Univ. of Detroit, Detroit, MI 48221

INTRODUCTION

Interpenetrating polymer networks (IPN's) are a novel type
of polyblend composed of crosslinked polymers. They are more or
less intimate mixtures of two or more distinct crosslinked polymers with no covalent bonds between the polymers. True IPN's
may be described as combinations of chemically dissimilar polymers
in which the chains of one polymer are completely entangled with
those of the other, i.e., a homogeneous morphology results.

Most of the IPN's reported to date have involved heterogeneous systems (1-9), usually one phase rubbery and the other glassy.
This combination of rubbery and glassy polymers often results in
a synergistic effect in properties (3,6,7), particularly mechanical properties. Either high impact plastics or reinforced rubbers result, depending on the morphology (i.e., phase continuity).
The only instances in which apparently homogeneous morphology has
resulted are with IPN's in which grafting between the component
polymer networks was a distinct possiblity (10-13).

This paper reports what we believe to be the first true IPN,
i.e., no grafting between polymers and a single phase morphology
(i.e., complete chain entanglement). In order to achieve this,
polymers of known compatibility were used. Thus, IPN's, pseudo-
IPN's (PDIPN's - only one polymer crosslinked), and linear blends
of polystyrene (PS), and poly(2,6-dimethyl-1,4-phenylene oxide)
(PPO) (whose compatibility has been reviewed elsewhere (14))
were prepared by the simultaneous interpenetrating network (SIN)
technique. The polystyrene was crosslinked by incorporating
divinylbenzene. Several methods have been reported to synthesize

203

crosslinked PPO using materials such as polysulfonazides and
hydroxymethyldiphenyl oxide (15,16).  However, these methods were
found to be difficult for simultaneous IPN's in which these two
polymer networks are formed in situ at about the same time.  In
this study, the PPO was brominated.  Crosslinking could then take
place with a tertiary amine.

The IPN's were studied by electron microscopy, ultimate property measurement, dynamic mechanical spectroscopy, and differential scanning calorimetry.

EXPERIMENTAL

Preparation

Brominated PPO:  The PPO was brominated by direct bromination
of polyphenylene oxide in solution.  One liter of s-tetrachloro-
ethane and 125.0g. (1.04 moles) of PPO (n = 0.55 dl./g) were combined in a 3 liter round bottom 3-necked flask equipped with
stirrer, thermometer, dropping funnel and condenser.  The reaction
mixture, maintained under a nitrogen atmosphere and illuminated
with a sunlamp, was stirred and heated to incipient reflux (136°C).
7.86 g. (0.0492M) of bromine was added dropwise.  Copious evolu-
tion of HBr carried off some of the bromine so another equal amount
of bromine was added.  The reaction was maintained at 136°C for
1 hour, then cooled and chloroform added.  The brominated PPO was
precipitated by the slow addition of methanol, filtered and washed
with methanol.  The material was dissolved in chloroform and
reprecipitated.  The collected material was broken up in a Waring
blender, filtered, washed well and dried in vacuo at 60°C over-
night.  A master batch of PPO (containing 4.75% of bromine) was
prepared by dissolving in toluene (25% by weight).

Linear Blends:  The PPO solution and styrene monomer (inhibitor removed) were mixed with 1% azobisisobutyronitrile (AIBN)
catalyst.  The mixture was poured between glass plates with a
Teflon spacer and subsequently polymerized at 70°C for 24 hours.
The glass plate mold was kept in a horizontal position so that an
even thickness sheet could be obtained.  Combinations of 75%, 50%
and 25% PPO by weight were made.

Pseudo-IPN's (PDIPN's):  Two types of pseudo-IPN's were prepared, one from crosslinked PPO (CPPO)/linear polystyrene (LPS)
(PDIPN-1) and the other from a linear PPO (LPPO)/crosslinked polystyrene (CPS (PDIPN-2).  The former PDIPN's were prepared from a
mixture of PPO solution with ethylene diamine (crosslinking agent)
and styrene monomer with AIBN.  1.4 g. of ethylene diamine was
added to 40 g. of PPO.  The mixture of PPO solution and styrene

monomer admixed with divinylbenzene and AIBN resulted in the
PDIPN-2.  The curing conditions and compositions were the same as
in the linear blends.

   Full IPN's (FIPN's):  FIPN's (both polymers crosslinked) were
prepared from the mixture of the PPO solution with ethylene diamine and styrene monomer with divinyl benzene and AIBN.  The curing conditions and compositions were the same as those for the
linear blends.

## Measurements

   Electron Micrographs:  The samples were prepared according
to Kato's osmium tetroxide staining technique (17) and a two-step
sectioning method.  The electron micrographs were taken by an AEI
6B and a Phillip's 300 transmitting electron microscope with a
magnification of 95,000.

   Differential Scanning Calorimetry:  The glass transition
temperatures ($T_g$'s) were determined on a Perkin-Elmer Differential Scanning Calorimeter DSC-2.  Measurements were carried out
from $300^{\circ}K$ to $500^{\circ}K$ under nitrogen at a scanning rate of $10^{\circ}C$ per
minute.  Specimen sizes were on the order of 20 mg.

   Ultimate Properties:  The tensile strengths were determined
on an Instron Tensile Tester at room temperature at a crosshead
speed of 2 in/min using dumbbell-shaped specimens (0.08" x
0.25" x 2").

   Rheovibron:  Dynamic viscoelastic properties were measured
on a Rheovibron Dynamic Viscoelastometer Model DDV-II (Toyo
Instrument Co.).  The measurements were taken over a temperature
range of $20^{\circ}C$ to $260^{\circ}C$ using a frequency of 110 Hz. and a heating
rate of $2^{\circ}C/min$.  Nitrogen gas was purged through the sample
chamber to avoid oxidation at high temperature.

## RESULTS AND DISCUSSION

## Crosslinking of PPO

   The ideal crosslinking reaction of brominated PPO is shown
below.  To confirm that crosslinking did indeed occur, solubility
studies and elemental analyses were performed.  The cured polymers
did not dissolve in either tetrahydrofuran or trichloroethane
(both good solvents for PPO); however, swelling did occur to about
twice the initial dimensions.  The elemental analyses (N, Br)*

*Carried out at Schwartzkopf Microanalytical Laboratory.

showed that only 10% of the bromine in PPO was involved in the
crosslinking reactions, the remaining bromine being inactive (most
likely aromatic bromine).

$$ 2 \quad \text{(aryl)}-CH_2Br \;+\; H_2NCH_2CH_2NH_2 $$

$$ \xrightarrow{\quad\quad} \quad \text{(aryl)}-CH_2-\overset{H}{N}CH_2CH_2\overset{H}{N}-CH_2-\text{(aryl)} \;+\; 2\;HBr $$

Based on the amount of bromine reacted, the number average molecular weight between two crosslink sites ($\overline{M}_c$) in PPO was 17,000.
The theoretical $\overline{M}_c$ of the crosslinked polystyrene was 7,260.

## Morphology

The morphology in all cases could clearly be identified by
electron microscopy and glass transition behavior.  The electron
micrographs of the component polymers (Figures 1 and 2) and the
blends (Figures 3 and 4) exhibit homogeneous phase mixing of the
PPO and the PS.  No fine structure could be seen at all.  The pure
PPO phase was darker (Fig. 1) due to the staining by osmium tetroxide and/or possibly due to the presence of bromine, while
the pure PS phase was not stained (white) (Fig. 2).  The electron-
micrographs in all cases did not show any domains of either
components.

The $T_g$'s of the FIPN's, PDIPN's and linear blends are listed
in Table I.  The two glass transition temperatures measured by
DSC and Rheovibron are quite close (within experimental error).
In all cases one single broad $T_g$ was observed, indicating complete
segmental mixing of the two polymers, in agreement with the electron microscopy.  The glass transition temperature in all cases
varied systematically with the composition.  The FIPN's showed
higher $T_g$ than the PDIPN's and the linear blends at the corresponding compositions.  This is undoubtedly due to the higher $T_g$'s
of the pure components (crosslinked) and does not relate to the

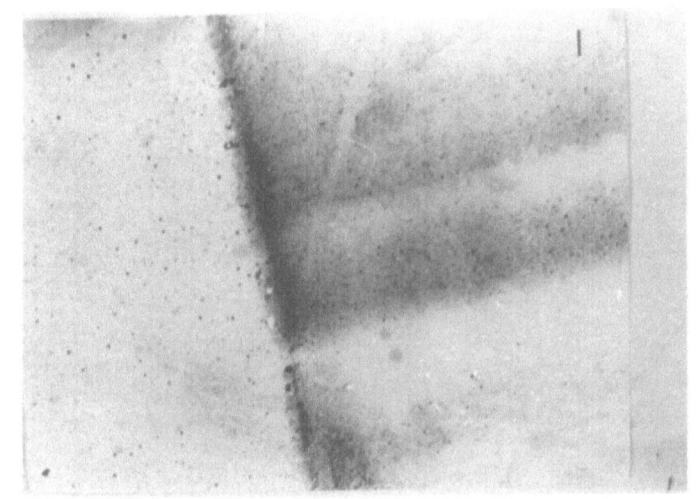

Fig. 2.   Electron Micrograph
of 100% Crosslinked
PS.

Fig. 1.   Electron Micrograph of 100%
Crosslinked PPO.

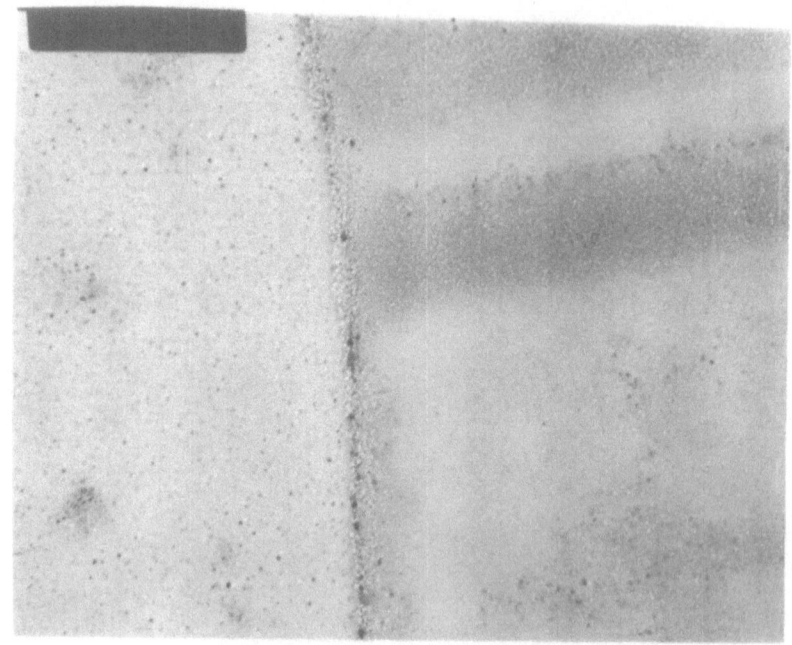

Fig. 4.   Electron Micrograph of
Linear Blend of 50%
LPPO/50% LPS.

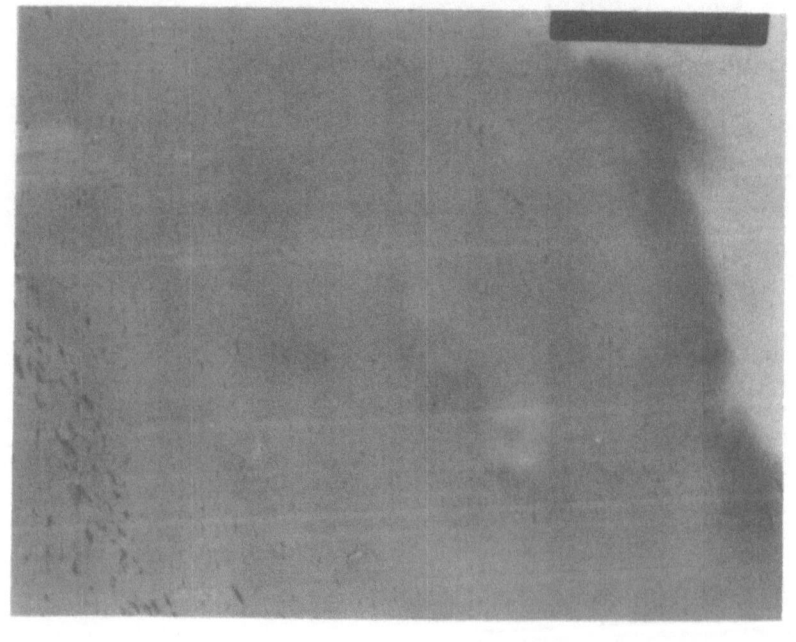

Fig. 3.   Electron Micrograph of FIPN
of 50% CPS/50% CPPO.

TABLE I.    GLASS TRANSITION TEMPERATURE ($T_g$) OF THE
PPO–PS BLENDS BY DSC AND RHEOVIBRON ($^{\circ}C$)

| Wt. % (PPO/PS) | 100/0 | 75/25 | 50/50 | 25/75 | 0/100 |
|---|---|---|---|---|---|
| FIPN's (CPPO/CPS) | 226 (232)* | 172 | 149 (145) | 120 | 83 (82) |
| PDIPN's-1 (CPPO/LPS) | 226 (232) | 266 | 234 (135) | 113 | 75 |
| PDIPN's-2 (LPPO/CPS) | 213 (222) | 162 | 138 (142) | 117 | 83 (82) |
| Linear Blends (LPPO/LPS) | 213 (222) | 152 | 130 (128) | 99 | 75 |

*Measured by Rheovibron.

C = crosslinked;  L = linear.

---

morphology differences.  The following empirical equation is often used to show the dependence of $T_g$ on the composition in random co-polymers and plasticized systems:

$$\frac{1}{T_g(cop)} = \frac{W_1}{T_{g1}} + \frac{W_2}{T_{g_2}} \qquad (1)$$

where $T_g(cop)$ is the observed glass transition of the copolymer and $W_i$ is the weight fraction of component i having a glass transition of $T_{gi}$.  A comparison of the predicted $T_g$'s using the above equation with the corresponding $T_g$'s measured shows the experimental values generally to be higher than the calculated values (see Table II). They are also generally lower than the $T_g$ average ($T_g(av)$ defined by:

$$T_g(av) = W_1 T_{g1} + W_2 T_{g2} \qquad (2)$$

which is in agreement with previous studies on another single phase IPN (13).  The amount of lowering of the $T_g$ may be expressed in terms of $\sigma$ (the full significance of $\sigma$ was discussed in Reference (13).

$$\frac{T_g - T_g(av)}{T_g(av)} \qquad \frac{-\sigma}{1 + \sigma}$$

TABLE II.   CALCULATED GLASS TRANSITION TEMPERATURE
OF THE PPO-PS FIPN's ($^\circ$C) (Using DSC
Values for Homopolymer $T_g$'s)

| Wt. % (PPO/PS) | 100/0 | 75/25 | 50/50 | 25/75 | 0/100 |
|---|---|---|---|---|---|
| $T_g$(cop) | 226 | 173.8 | 142.6 | 110.5 | 83 |
| $T_g$(av) | 226 | 190.3 | 154.5 | 118.8 | 83 |
| $T_g$ (measured) | 226 | 172 | 149 | 120 | 83 |
|  | – | 0.041 | 0.012 | -0.003 | – |

If significant phase mixing occurs, σ would be expected to be
positive or zero (i.e., $T_g$ ≤ $T_g$(av) as it was (effectively) for
these IPN's.  For an IPN, σ is expected to reach a maximum very
close to the maximum in crosslink density (which includes physical
entanglement crosslinks resulting from the superposition of the two
networks) (17).  The maximum value of σ is at an IPN of 75% PPO/
25% PS, suggesting, therefore, that the maximum amount of interpenetration occurs at this point, since interpenetration would be
expected to manifest itself in additional chain entanglements.

## Tensile Strength

The tensile strengths of the IPN's, PDIPN's and linear blends
as a function of composition are shown in Fig. 5.  The FIPN's
showed an enhancement in tensile strength with a maximum at 25%
PS (75% PPO).  This behavior is very similar to the polyblends of
Fried et al (18), except that a minimum at about 25% PPO did not
occur.  This maximum occurred at the same composition as the maximum in σ (greatest $T_g$ lowering), further substantiating our previous hypotheses which describe the enhancement to increased chain
entanglements.

The PDIPN's and linear blends do not show a maximum, contrary
to the results of Fried et al (18), but in general do show a minimum.  In addition, the tensile strengths of the FIPN's are higher
than those of the PDIPN's and the linear blends at the same composition.  There is no significant difference in the tensile strength,
in general, between the PDIPN's and linear blends.  The enhancement
in tensile strength is most likely due to the permanent chain entanglements in the FIPN's.  In the case of PDIPN's and linear blends,
the entanglements are only temporary and "untangling" can occur at
large deformations because one or both polymers are linear (topological constraints do not exist).

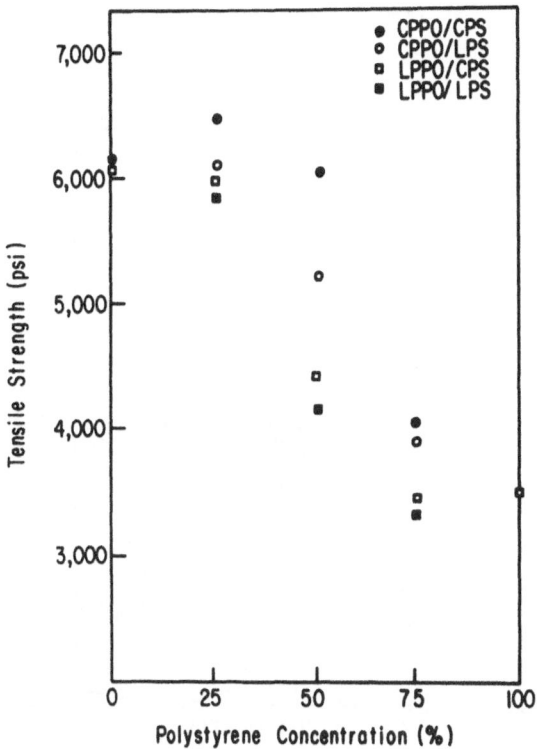

Fig. 5.   Tensile Strength vs. Polystyrene Concentration.

## Dynamic Mechanical Properties

The storage modulus, E', of FIPN, PDIPN and linear blends with 50% PS as a function of temperature are shown in Fig. 6. From room temperature up to the respective $T_g$'s, the moduli of all the IPN's are higher than those of the pure components with the modulus of the FIPN being the greatest. This synergism in modulus has been noted before in compatible PPO/PS blends (19) and has been described to increase in packing density due to blending. One would, therefore, expect the FIPN's to exhibit the greatest increase in modulus, since there should be a greater amount of permanent chain entanglements between the two networks, which would result in a greater increase in packing density. The moduli sharply decrease near the $T_g$'s of each blend, respectively, as usual. The loss modulus, E'', of this sample as a function of temperature is also shown in Fig. 6. The loss modulus data shows a single $T_g$, indicating extensive phase mixing. The $T_g$'s, however, are broader than those of the pure components.

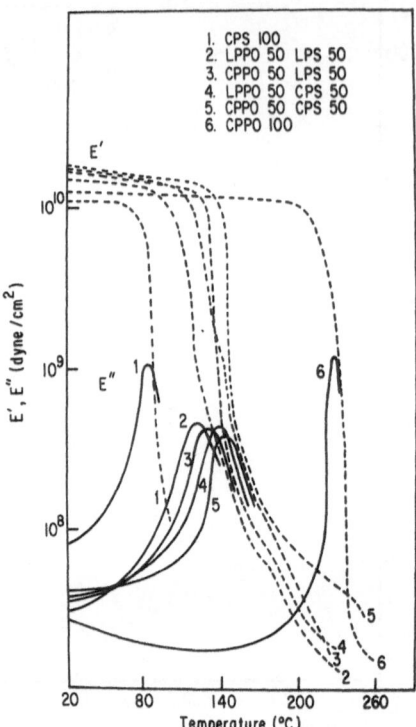

Fig. 6.    Temperature Dependence of
           Storage (E') and Loss (E")
           Moduli.

CONCLUSIONS

    A.  FIPN's, PDIPN's and linear blends of poly(2,6-dimethyl-
1,4-phenylene oxide) and polystyrene all exhibited single phase
behavior as evidenced by glass transition analysis and electron
microscopy.  Thus, for the first time, true interpenetrating polymer networks have been produced, i.e., homogeneous morphology with
little or no possibilities of covalent bonds between the component
polymers.

    B.  The glass transition temperatures of the FIPN's agree with
our previous theory on the position of the $T_g$ of IPN's, i.e., the
$T_g$'s were, in general, lower than the $T_g$(av).  The amount of lowering corresponded with the maximum in tensile strength, again in
agreement with our theory.

    C.  An enhancement in tensile strength occurred with the
FIPN's, but not with the PDIPN's or linear blends, verifying the
role of interpenetration in mechanical properties.

D. A similar increase in modulus occurred with all the IPN's, the enhancement being greatest with the FIPN's, again verifying the role of interpenetration in mechanical properties.

## ACKNOWLEDGEMENTS

The authors would like to acknowledge with gratitude the support of the National Science Foundation under NSF Contract DMR-7805938, The American Chemical Society under Petroleum Research Fund Grant 11504-AC7, and the help of Drs. A. Hay and D. M. White and Mrs. B. M. Boulette of the General Electric Research Center for their synthesis of the brominated PPO.

## REFERENCES

1. M. Matsuo, T. K. Kwei, D. Klempner, H. L. Frisch, Polymer Eng. Sci. 10, 327 (1970).
2. L. H. Sperling, D. A. Thomas II, V. Huelck, Macromolecules 5, 340 (1972).
3. A. J. Curtius, M. J. Covitch, D. A. Thomas, L. H. Sperling, Polymer Eng. Sci. 12, 101 (1972).
4. D. Klempner, H. L. Frisch, K. C. Frisch, J. Polym. Sci. A-2 8, 921 (1970).
5. D. Klempner, H. L. Frisch, K. C. Frisch, J. Elastoplastics 3, 2 (1971).
6. S. C. Kim, D. Klempner, K. C. Frisch, H. L. Frisch, H. Ghiradella, Polym. Eng. Sci. 15, 339 (1975).
7. S. C. Kim, D. Klempner, K. C. Frisch, H. L. Frisch, J. Appl. Polym. Sci. 21, 1289 (1977).
8. S. C. Kim, D. Klempner, K. C. Frisch, H. L. Frisch, W. Radigan, Macromolecules 9, 258 (1976).
9. S. C. Kim, D. Klempner, K. C. Frisch, H. L. Frisch, Macromolecules 9, 263 (1976).
10. K. C. Frisch, D. Klempner, S. Migdal, H. L. Frisch, H. Ghiradella, Polym. Eng. Sci. 14, 76 (1974).
11. K. C. Frisch, D. Klempner, S. Migdal, H. L. Frisch, J. Polym. Sci. A-1 12, 885 (1974).
12. K. C. Frisch, D. Klempner, S. Migdal, H. L. Frisch, J. Appl. Polym. Sci. 19, 1893 (1975).
13. H. L. Frisch, K. C. Frisch, D. Klempner, Polymer Eng. & Sci., 14, 646 (1974).
14. W. J. MacKnight, F. E. Karasz, J. R. Fried, "Polymer Blends," (D. R. Paul and S. Newman, eds.), Vol. I, 185, Academic Press, New York, 1978.
15. E. E. Bostick and A. R. Gilbert, Fr. Pat. No. 1,518,441, 22, Mar., 1968.
16. S. Schmukler, U. S. Pat. No. 3,396,146, 06, Aug., 1968.

17.  K. Kato, Polymer Eng. & Sci., 7, 38 (1967).
18.  J. R. Fried, W. J. MacKnight, F. E. Karasz, Coatings and
     Plastic Preprints, Honolulu A.C.S. Meeting (1979).
19.  L. W. Kleiner, F. E. Karasz, W. J. MacKnight, SPE 36th Annual
     Technical Conference, Washington, D. C., April 24, 1978,
     pp. 243-248.

COMPATIBILITY AND TENSILE PROPERTIES OF PPO[*] BLENDS[+]

J. R. Fried

Department of Chemical and Nuclear Engineering
and the Polymer Research Center
University of Cincinnati
Cincinnati, Ohio 45221

W. J. MacKnight and F. E. Karasz
Polymer Science and Engineering Department
and the Materials Research Laboratory
University of Massachusetts
Amherst, Ma. 01003

ABSTRACT

    Young's modulus, yield (break) strength, and elongation to
yield (break) have been measured for blends of poly (2,6-dimethyl-1,
4-phenylene oxide)(PPO) with polystyrene (PS), poly (p-chlorostyrene)(PpClS), and random copolymers of styrene and p-chlorostyrene)(pClS). The significant difference between blend compositions is the compatibility of PPO with each styrene polymer.
Blends of PPO with PS or copolymers with 67.1 mole % or less pClS
are compatible (i.e. one Tg) and show small synergistic maxima in
modulus, strength, and elongation as a function of PPO composition.
These maxima correspond to observed maxima in packing density as a
result of specific interactions contributing to blend compatibility.

---

[*]Registered trademark of the General Electric Company

[+]This paper was originally published in the Journal of Applied
Physics, Vol. 50, No. 10, Part II under the title "Modeling of
Tensile Properties of Polymer Blends: PPO/Poly(styrene-co-p-
chlorostyrene". © 1979 American Institute of Physics.
Reprinted with permission.

215

A rule of mixture for one phase systems with an adjustable compatibility parameter gives adequate fit to the observed composition
dependence of the modulus.

In a narrow composition range between 67.8 and 68.6 mole %
pClS, copolymers exhibit partial miscibility with PPO.  Two mixed
composition phases are present.  Moduli of these transitional
blends follow the same form of synergistic dependence on blend
composition as do the compatible blends but strength and elongation
exhibit a sigmoidal relation to blend PPO content.  At about 20%
PPO, strength (and elongation) reach a minimum as predicted by a
simple composite-model for a dispersed phase with good adhesion to
the matrix.  A maximum is reached at ca.  80% PPO at which composition blend test specimens yield prior to failure.

Blends of PPO with PpClS and with copolymers of > 68.6 mole %
pClS exhibit a broader minimum in strength (and elongation) but a
similar maximum at 80% PPO.  Unlike the compatible and transitional
blends, moduli follow a nonsynergistic composition dependence
adequately represented by the series model for two phase systems.

INTRODUCTION

The mechanical properties of filled polymer composites have
been widely studied and theories modeling their behavior have been
successful in achieving at least a practical understanding of how
filler shape, size, concentration, modulus, and interfacial adhesion
relate to the properties of the matrix polymer in determining the
overall mechanical response of the composite.  Much less attention
has been directed toward understanding how the mechanical properties of polyblends, in which both components are polymeric, are
related to the properties of the individual components.  Qualitatively, it is recognized that blends of two polymers that are not
mutually miscible, i.e. are incompatible, form separate phases that
are generally less well-defined in size and shape than in controlled
composite formulations.  Films, fibers, and molded parts of such
incompatible blends are opaque and have low strength and toughness
as a result of poor adhesion between phases (1).  At the other
extreme, compatible polyblends, in which the component polymers do
not phase separate but instead form a microscopically homogeneous
single phase, have high strength and form clear films.  In certain
cases, both modulus (2) and tensile strength (3) have been reported
to be greater at particular blend compositions than the corresponding properties of either polymer in the unblended state.  Such
synergism in modulus and strength is apparently achieved at the
expense of impact strength as ductile polymers will undergo embrittlement as a result of blending with another compatible polymer in
apparent analogy to polymers antiplasticized with low molecular
weight additives (3,4).

   In this paper, an attempt is made to relate the mechanical
(tensile) properties of a family of related polyblends to the state
of compatibility of the blend.   The prototype compatible blend
studied is that of poly (2,6-dimethyl-1, 4-phenylene oxide)(PPO)
and polystyrene (PS). Evidence for the compatibility of PPO and PS
is substantial and is reviewed elsewhere (5).   Films molded from
blends of PPO and PS are optically clear and exhibit a single
composition dependent glass transition temperature (Tg).

   By contrast, PPO is not compatible with chlorinated PS,
either poly(p-chlorostyrene)(PpClS) or poly(o-chlorostyrene)(PoClS)
(6).   Blends of PPO with either PpClS or PoClS form opaque films and
exhibit two glass transitions identical in temperature and dispersion width to those of the corresponding unblended polymers.   The
independence of phase Tg on blend composition (weight fraction PPO)
for PpClS/PPO blends is illustrated in Fig. 1.   In addition, fracture replicas of PpClS/PPO blends indicate macroscopic phase separation of the component polymers into large irregular-shaped
domains(7).

   Blend compatibility can be varied systematically by blending
PPO with random copolymers of styrene and p-chlorostyrene (pClS)
(6,8).   Copolymers with low pClS content ($\leq$ 67.1 mole % pClS) are
compatible with PPO as evidenced by film clarity and by a single Tg
at each blend composition.   Recently, Couchman (9) has shown that
the Tg of these compatible blends follows a composition dependence
given as

$$\ln(Tg/Tg_1) = [W_2 \Delta Cp_2 \ln(Tg_2/Tg_1)]/[W_1 \Delta Cp_1 + W_2 \Delta Cp_2] \qquad (1)$$

where W and $\Delta C_p$ are weight fraction and change in heat capacity at
Tg, respectively, for components 1(copolymer) and 2(PPO) as indicated.   The form of this dependence is illustrated for blends of a
copolymer with 58.5 mole % pClS   (Copolymer B) and PPO in Fig. 1.
As PPO is blended with copolymers of increasing pClS content, the
dispersion width of the single glass transition of the compatible
copolymer/PPO blends increases.   This broadening of the transition
width has been directly related to an increase in localized concentration fluctuations in the blend, i.e. increasing blend heterogeneity (10).

   Blends of PPO and copolymers in the narrow composition range
between 67.8 and 68.6 mole % pClS exhibit phase separation as
evidenced by the detection of two glass transitions; however, these
transitions are not as well defined as those of the incompatible
PpClS/PPO, PoClS/PPO, and higher pClS copolymer/PPO blends.   They
are considerably broadened, diminished in intensity ($\Delta C_p$), and
shifted in temperature when compared to the corresponding unblended
components.   This intermediate character of these "transitional"
blends is illustrated by the Tg-composition dependence of

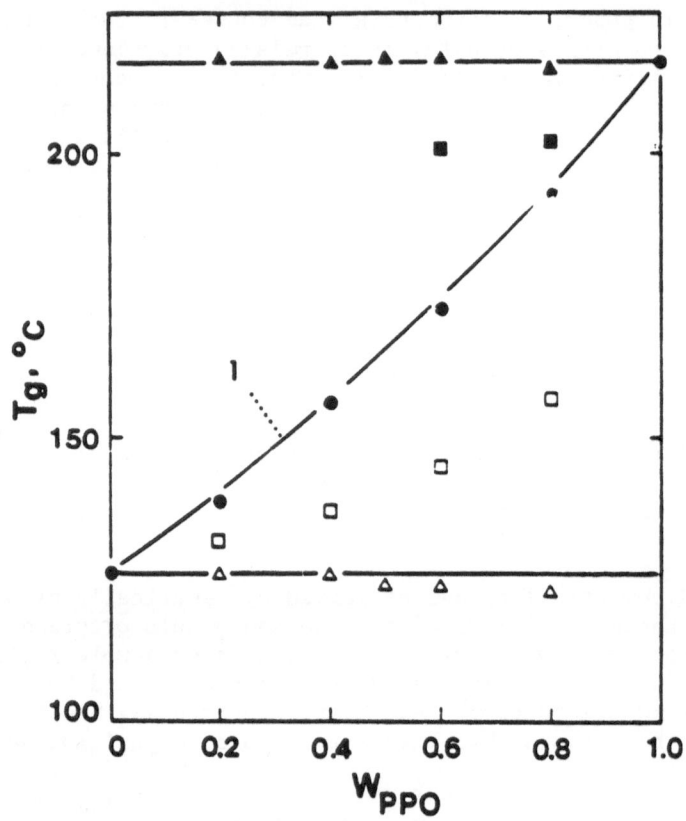

Fig. 1   Blend glass transition temperature (Tg) as a function of
         weight fraction PPO ($W_{PPO}$):  (●), Copolymer B/PPO;
         ( ■ ), Copolymer D/PPO, high temperature transition;
         ( □ ), Copolymer D/PPO, low temperature transition;
         ( ▲ ), PpClS/PPO, high temperature transition;
         ( △ ), PpClS/PPO, low temperature transition (each point
         dropped 7°C on the ordinate for purpose of comparison
         with the lower Tg copolymers).  Curve 1 was drawn from
         values of Tg's calculated by means of the Couchman equation
         (Eq. 1) for the compatible Copolymer B/PPO blends.

Copolymer D (67.8 mole % pClS)/PPO blends in Fig. 1 where Tg data
points fall between the curve given by Eq. 1 for the compatible
blends and the horizontal lines representing the composition independent Tg's of the two-phase incompatible blends.

     The shifting of component Tg's suggests that the above mar-
ginally compatible blends are composed of mixed-composition domains,
a PPO-rich (high Tg) phase and a copolymer-rich (low Tg) phase.  As
has been observed directly for other incompatible blends (11),

it is probable that the domains themselves are separated by diffuse interfacial regions. The presence of large, diffuse interfacial regions would explain the observed reduction of blend $\Delta C_p$ (6). Films of these blends are either clear or hazy depending on the PPO composition of the blend.

In summary of the above results, PPO/poly(styrene-co-p-chloro-styrene) blends may be divided into three compatibility categories depending upon the pClS composition of the copolymer as follows:

| Mole % pClS | Classification | Blend Morphology |
|---|---|---|
| 0-67.1 | compatible | homogeneous |
| 67.8-68.6 | transitional | small mixed composition domains; large interfacial regions |
| 75.4-100 | incompatible | large homogeneous domains; small interfacial regions |

The mechanical properties of blends representing these three blend categories have been measured as a function of blend composition (volume fraction PPO) and are reviewed below with respect to current theories of composite behavior.

EXPERIMENTAL

Methods of copolymer and blend preparation have been detailed elsewhere (6). The physical and mechanical properties of those polymers whose blend mechanical properties are reported here are summarized for convenience in Table I.

Minature dumbbell-shaped specimens of each polymer and blend were molded from pieces of compression molded films by means of a Mini-Max Injection Molder (Custom Scientific Instruments) whose operation and mixing characteristics have been reviewed by Maxwell (12). Gauge length of the molded dumbbells was 8.9 mm at a cross-sectional diameter of 0.157 cm.

Cup and mold temperatures were raised by increments of 20°C for each blend increment of 20 weight % PPO from a lower limit of 250°C for PS, the copolymers, PpClS, and PoClS to a high of 340°C for unblended PPO. After injection, molds were removed, placed on a large metal plate, and allowed to slowly cool to ambient temperature (23°C). In this manner, blends were cooled at rates comparable to the controlled rates used in previous DSC studies of these blends (6). Studies of n-hexane induced crazing of PS/PPO molded specimens using the above techniques indicated no preferential orientation of crazes whereas rapidly quenched samples

TABLE I

Properties of Blend Polymers

| Polymer | mole % pClS | $T_g$ (°C)* | $\bar{M}_n \times 10^{-3}$ | $\bar{M}_w \times 10^{-3}$ | E (GPa) | σ (MPa) | ε (%) |
|---------|-------------|-------------|----------------------------|----------------------------|---------|---------|-------|
| Copolymer B | 58.5 | 125 | 95 | 208 | 3.40 | 49[b] | 1.2[b] |
| Copolymer D | 67.8 | 126 | 100 | 192 | 3.15 | 45[b] | 1.1[b] |
| PpClS | 100 | 132 | 128 | 217 | 3.49 | 46[b] | 1.1[b] |
| PPO | ---- | 216 | 17 | 35 | 2.66 | 71[y] | 2.7[y] |

*DSC; 20°/min
b, break
y, yield

exhibited craze orientation parallel to the injection (axial) direction (2b).

Measurements of tensile properties were made at room temperature by means of a Tensilon UTM-II mechanical tester (Toyo Baldwin Co., Ltd.) at a constant crosshead speed of $0.2$ mm min$^{-1}$ corresponding to a nominal strain rate of $3.75 \times 10^{-4}$ sec$^{-1}$. Young's modulus (E) was arbitrarily defined as the secant modulus at 0.6% elongation and is expressed in SI units of GPa (GPa = $10^{10}$ dynes/ cm$^2$). Ultimate strength ($\sigma_u$) and elongation ($\varepsilon_u$) were defined as the stress and engineering strain, respectively, at break. Yield strength ($\sigma_y$) and elongation ($\varepsilon_y$) were taken as the stress and engineering strain at the maximum in the stress-strain curve. The values of $\sigma$ and $\varepsilon$ are expressed in MPa (MPa = $10^7$ dynes/cm$^2$) and %, respectively. Measured $\varepsilon$ was corrected for finite sample gauge length and instrumental compliance as detailed elsewhere (13). Values for E, $\sigma$, and $\varepsilon$ were reported as population mean values for 4-15 samples according to standard small sampling techniques; error bars indicate 95% confidence limits.

RESULTS

Representative stress-strain curves for the compatible Copolymer B/PPO, transitional Copolymer D/PPO, and incompatible PpClS/PPO blends are illustrated in Figs. 2-4. Qualitatively, one of the more notable results illustrated by these series of curves is an apparent embrittlement at high PPO blend compositions as the mode of failure changes from one of predominantly brittle fracture at below 80% PPO to ductile or yielding behavior at and above 80% PPO. Embrittlement of PPO by PS has been reported to accompany a suppression of the low temperature $\beta$ relaxation of PPO in analogy to the observed embrittlement of ductile polymers antiplasticized by low molecular weight additives (3,4). Wellinghoff and Baer (3) have shown that this observed embrittlement of PPO by PS in these compatible blends corresponds to a change in the process of deformation from one of diffuse shear banding (Type II glass) to one of extensive craze initiation and growth (Type I glass). The deformation microstructure of the copolymer/PPO blends, although not reviewed in the present study, are under investigation and will await future publication.

As Fig. 2 illustrates, the stress at break (or yield) of the compatible Copolymer B blends rises to a maximum at between 60 and 80% PPO. This apparent enhancement in strength has been observed in the case of polymers antiplasticized by low molecular weight additives (14) and suggests that the compatible copolymers act as polymeric antiplasticizers for PPO. This analogy to antiplasticization has been applied to other compatible polymer blends (15,16). Similar synergistic dependence of tensile strength on blend composition for PS/PPO blends has been reported (13,17). In a study of

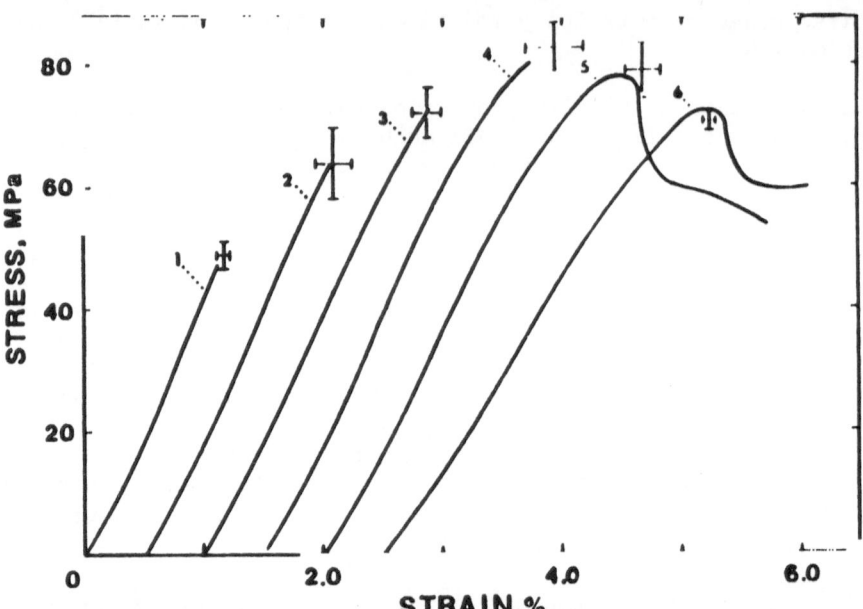

Fig. 2  Representative stress–strain curves of the compatible
Copolymer B/PPO blends.  Curve 1, Copolymer B; 2, 20% PPO;
3, 40% PPO; 4, 60% PPO; 5, 80% PPO; 6, 100% PPO.  Error
bars indicate 95% confidence intervals for both stress
and strain at break (or yield) for sample populations.
Curves 2–6 have been sequentially shifted 0.5% in strain
along the abscissa for purpose of comparison.

tensile behavior over several decades of strain rate, Yee (3) has
shown that actually two maxima in strength may be evident in the
case of PS/PPO blends:  one at below 20% PPO and the other at above
75% PPO.  The low PPO maximum was found to decrease in intensity
with decreasing strain rate ($\dot{\varepsilon}$).  A second low % PPO maximum may
occur for the compatible PS/PPO and Copolymer B/PPO blends reported
here, but Yee's results suggest that for the particular blend
compositions (none less than 20% PPO) and relatively large strain
rates chosen here, this maximum may escape detection.

Stress–strain curves for the transitional Copolymer D/PPO and
incompatible PpClS/PPO blends given in Figs. 3 and 4, respectively,
again indicate embrittlement at 60 to 80% PPO.  Unlike unblended
PPO and the high PPO content compatible blends that yield and cold
draw, the high PPO content two-phase blends do not appear to initiate a stable neck region and failure occurs shortly after the
yield point.

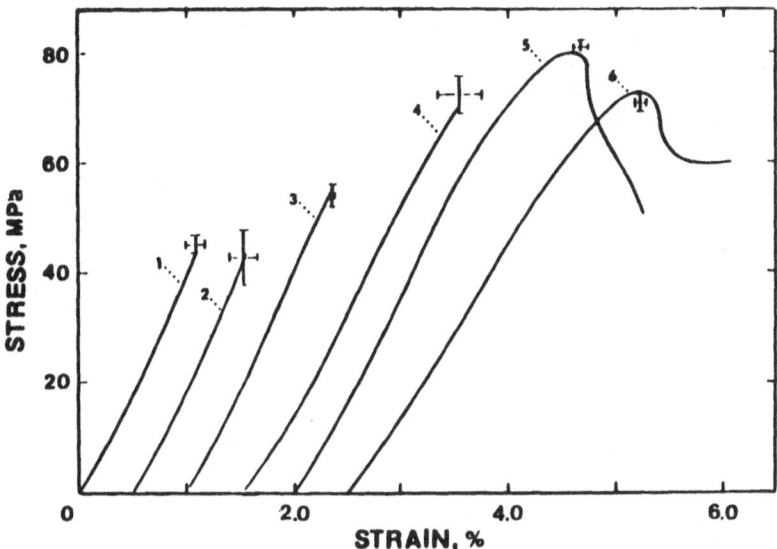

Fig. 3   Stress-strain curves of the transitional Copolymer D/PPO
         blends.   Identification is the same as in Fig. 2 except
         that Curve 1 represents unblended Copolymer D (0% PPO).

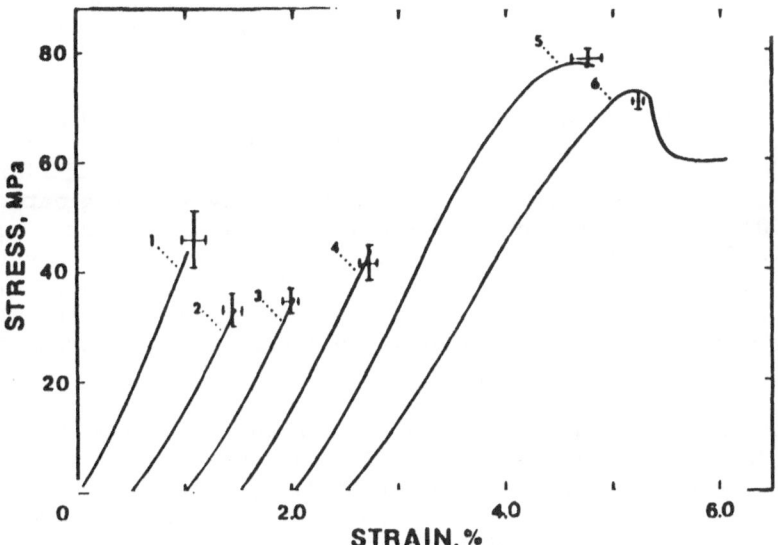

Fig. 4   Stress-strain curves of the incompatible PpClS/PPO blends.
         Identification is the same as in Fig. 2 except that
         Curve 1 represents unblended PpClS (0% PPO).

In comparison to the monotonic increase in strength exhibited by the compatible Copolymer B blends with increasing PPO content, the transitional Copolymer D and incompatible PpClS blends (Figs. 3 and 4) reach a minimum in break strength at 20 to 40% PPO. For example, the break strength of 80% PpClS/20% PPO is approximately 28% lower than the break strength of PpClS alone. These observations of break strength reduction along with the results of blend modulus behavior are analyzed below with respect to current theories of polymer composite behavior.

DISCUSSION

Modulus

The maximum modulus of a composite is given by the rule of mixture (18).

$$E = V_1E_1 + V_2E_2 \qquad\qquad (2)$$

where V represents the volume fraction of component 1 or 2. The lowest value of modulus is given by the series model (19).

$$1/E = V_1/E_1 + V_2/E_2 \qquad\qquad (3)$$

Values predicted by other equations such as those proposed by Kerner, Nielsen, Van der Poel, Grezczuk, Sato and Furukawa, Eilers-Van Dijck, and others fall between the bounds set by Eqs. 2 and 3.

Until recently, little attention has been given to predicting the modulus of polyblends. The first obvious distinction between polyblends and composites is that whereas the ratio of filler modulus to polymer modulus in composites is typically greater than 20, (20) the ratio of moduli for polyblends is very nearly equal to unity. For this reason of component modulus equivalency in polyblends, the form of the dependence of polyblend moduli on blend composition may be difficult to model in any meaningful manner within the limitation of typical scatter of experimental data.

Recently, Kleiner et al (2) have shown that the moduli of the compatible PS/PPO blends fall outside the upper bounds given by Eq. 2. Instead of the classical composite results, the blend moduli are reported to follow a composition dependency given by the general equation cited by Nielsen (21) for one-phase binary mixtures in the specific form given by Kleiner

$$E = V_1E_1 + V_2E_2 + \beta_{12}E_1E_2 \qquad\qquad (4)$$

The empirical interaction term, $\beta_{12}$, in Eq. 4 is given as

$$\beta_{12} = 4E_{12} - 2E_1 - 2E_2 \qquad (5)$$

where $E_{12}$ represents the measured modulus of a 50/50 (PS/PPO) blend. As an interaction term, $\beta_{12}$ may be a relative measure of blend compatibility. It was shown that $\beta_{12}$ increased with decreasing molecular weight of the PS component, _i.e._ in the direction of increasing blend compatibility. Kleiner postulated that the origin of the synergism in modulus suggested by the form of Eq. 4 and demonstrated experimentally in the case of PS/PPO is the observed increase in packing density due to blending. Spectroscopic evidence for specific interactions between PPO and PS that may account for such densification has been given (22).

The calculated moduli of the compatible Copolymer B/PPO blends are plotted against volume fraction PPO ($V_{PPO}$) in Fig. 5. The solid curve is drawn using Eq. 4 and an empirical value of 0.66 for $\beta_{12}$ as found by Kleiner to give best fit for the PS/PPO blends when PS is Monsanto HH 101 grade (similar molecular weight and molecular weight distribution to the copolymer). The good

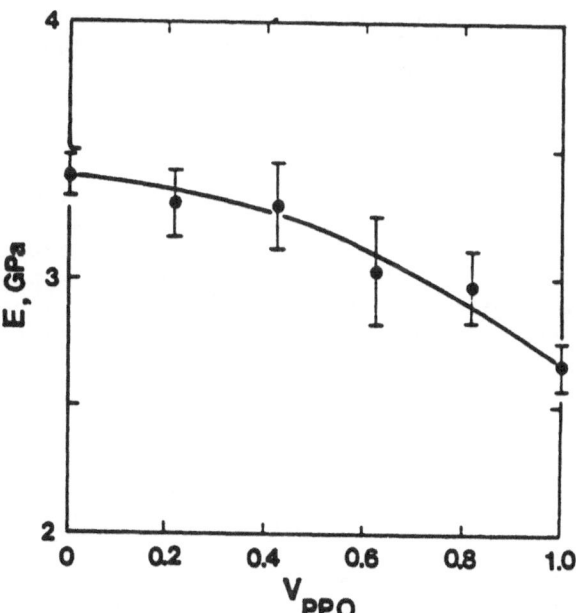

Fig. 5   Young's modulus (E) as a function of volume fraction PPO ($V_{PPO}$) for the compatible Copolymer B/PPO blends. The curve was drawn by use of the modified rule of mixtures for composites (Eq. 4).

agreement between the experimental moduli and the empirical Kleiner curve as illustrated in Fig. 5 suggests an equivalent compatible state in the Copolymer B/PPO blends as compared to the PS/PPO blends and as substantiated in prior publications by calorimetric (6) and dielectric (10) studies.  Scanning electron micrographs of cold fractured tensile specimens of these compatible copolymer/PPO blends reveal no evidence of macrophase structure at 1000X (13).

The transitional blends as defined in the prior section also appear to follow Eq. 4 as illustrated by the data and the empirical curve in Fig. 6.  It is noted that both the compatible and transitional blends exhibit packing densification (6) supporting the explanation given by Kleiner for the modulus behavior.

Finally, moduli of the incompatible PpClS/PPO blends are plotted against blend composition in Fig. 7.  As shown, experimental values are much lower on the ordinate than would be predicted for a compatible (and transitional) blend on the basis of Eq. 4 (upper curve in Fig. 7) but are well approximated by a composite series model given by Eq. 3 (lower curve).[*]

Tkacik (7b) has shown by means of transmission electron micrographs of molded films of PpClS/PPO blends fractured at above the Tg of PpClS that these blends exhibit macrophase separation (> 10μm) as illustrated by the series of micrographs in Fig. 8. At blend compositions of 25 and 50% PPO, these blends are characterized by large irregular shaped domains of PPO polymer dispersed in a PpClS matrix.  Phase inversion resulting in a PpClS dispersed phase and PPO matrix occurs at 75% PPO blend content. It is noted that the specific phase morphology of the injected molded tensile specimens used in the present study may differ from the compression molded film morphology.  In the case of extruded samples of polymer blends displaying macrophase separation, Van Oene (25) indicates that the dispersed phase may appear as either ribbons (stratification) or droplets independent of shear strain rate but dependent of the post–extrusion thermal history.  A study of the effect of morphology and phase inversion on the mechanical properties of the incompatible PPO blends is presently in progress.

---

[*]The moduli of PPO/glass bead composites are reported by Trachte and DiBenedetto (23) to follow the Kerner equation and by Wambach et al (24), to follow the equation of Van der Poel.  These may therefore be more appropriate choices in modeling the composite behavior of the PpClS/PPO blends but considering the low moduli ratio of the PpClS/PPO blends, $E_{PpClS}/E_{PPO} = 1.3$, the difference between values of moduli predicted by use of either of these equations or use of the simpler Eq. 3 would be slight.

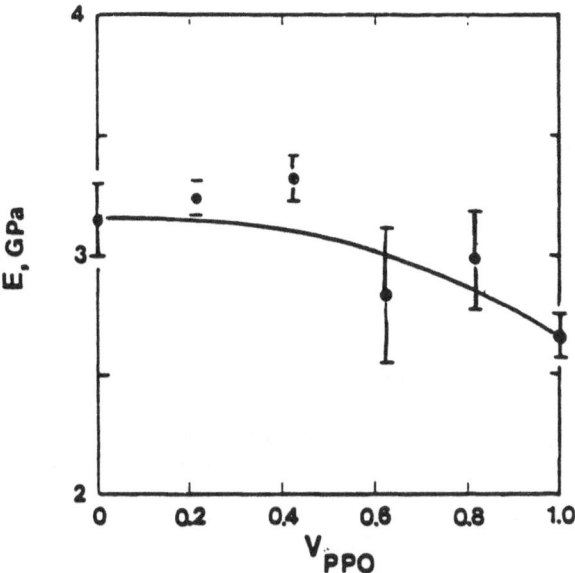

Fig. 6   Young's modulus versus volume fraction PPO for the
transitional Copolymer D/PPO blends.   Curve, Eq. 4.

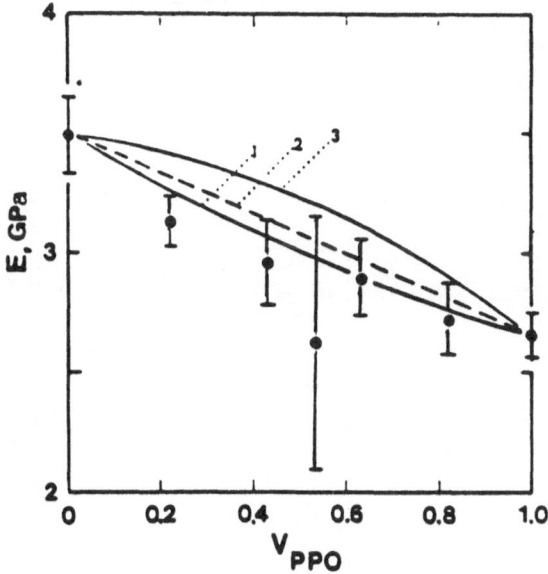

Fig. 7   Young's modulus versus volume fraction PPO for the incompatible PpClS/PPO blends.   Curve 1, the series model
(Eq. 3); curve 2, rule of mixtures (Eq. 2); curve 3,
modified rule of mixtures (Eq. 4).

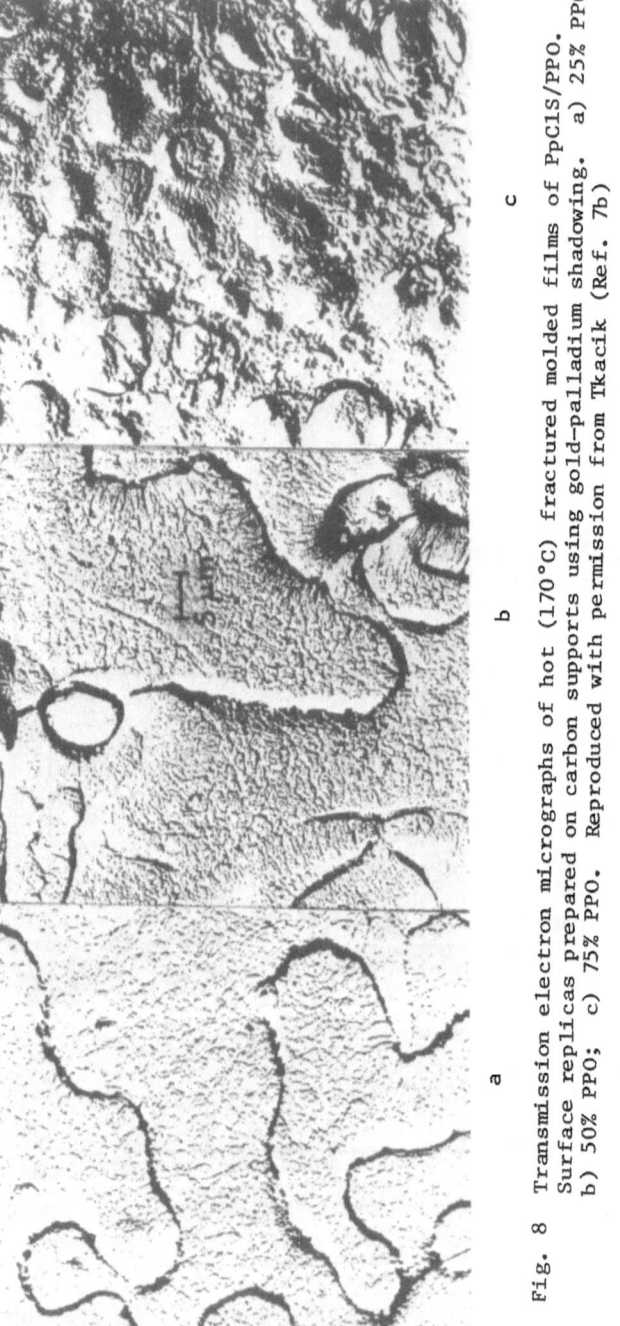

Fig. 8   Transmission electron micrographs of hot (170°C) fractured molded films of PpClS/PPO. Surface replicas prepared on carbon supports using gold-palladium shadowing.   a) 25% PPO; b) 50% PPO;   c) 75% PPO.   Reproduced with permission from Tkacik (Ref. 7b)

In addition to the composite-like morphology of the PpClS/PPO blends, these incompatible blends exhibit no blend densification in contrast to the compatible and transitional blends (6). Absence of blend densification would suggest no synergism in modulus in accordance with the Kleiner argument and in agreement with the experimental results.

## Tensile Strength

Strength at break (or yield) is plotted against volume fraction PPO ($V_{PPO}$) in Figs. 9-11 for the compatible, transitional, and incompatible blends whose moduli are plotted in Figs. 5-7. As previously reported for the prototype PS/PPO blends, (3,13) and as qualitatively shown by the stress-strain curves in Fig. 2, the compatible Copolymer B/PPO blends (Fig. 9) exhibit apparent synergism in tensile strength; i.e., the highest blend break

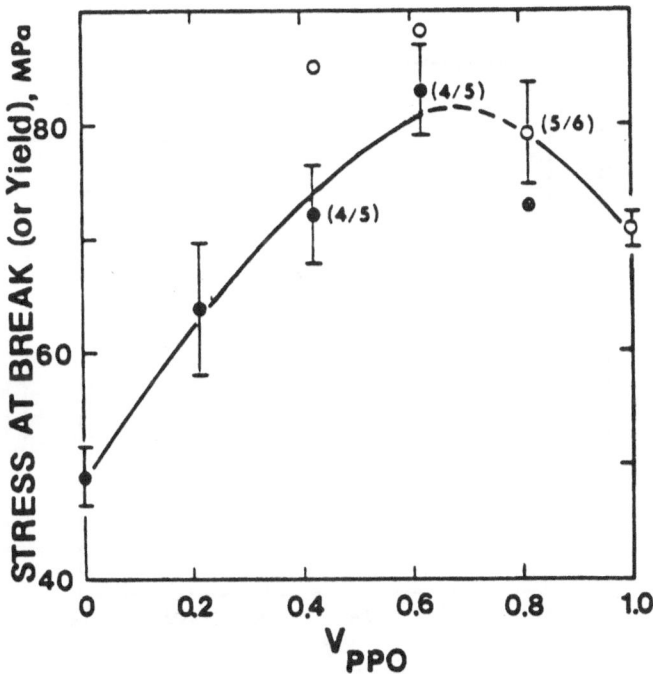

Fig. 9   Tensile stress at break (or yield) versus volume fraction PPO for the compatible Copolymer B/PPO blends. ( ● ), break strength; ( ○ ), yield stress. Error bars indicate 95% confidence limits. Values within parenthesis indicate fraction of samples within test populations that fail by predominant mode in the embrittlement region.

strength is greater than the stress at break of the copolymer alone
and correspondingly the highest yield stress of the blend is
larger that the measured yield stress of PPO in the unblended state.
Embrittlement, or the transition from the brittle to ductile mode
of failure, occurs at the same 60 to 80% PPO range in blend
composition as observed in the PS/PPO blends (3,13).  As indicated
in Figs. 9-11, some tensile specimens break by either brittle or
ductile failure at the same blend composition.  The fraction of
total specimens tested that fail in the predominant mode is indicated within the parentheses in each figure.  In all cases, the
yield stress is greater than the break stress.  These observations
may be explained in terms of the failure criteria proposed by
Nicolais and DiBenedetto (26) for which brittle failure will occur
if an individual sample defect grows to a critical defect size
before the stress-strain curve reaches a maximum.  Variations in
defect size and defect size distribution within a tensile sample
population result in a proportion of some samples failing in the
brittle mode and some in the ductile mode within the embrittlement
region.

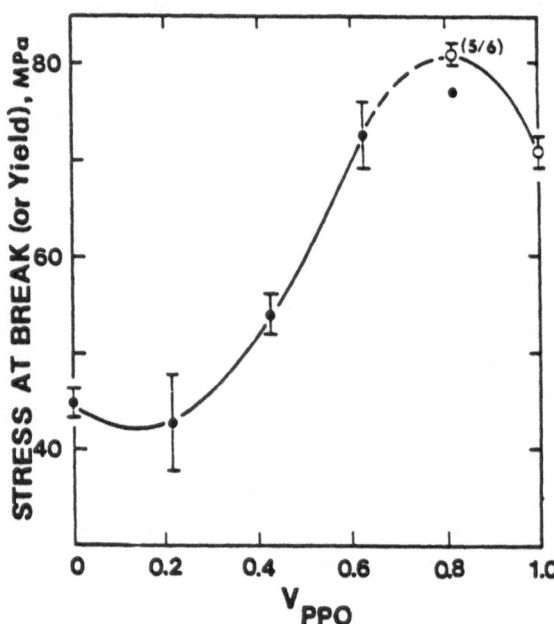

Fig. 10   Tensile stress at break (or yield) versus volume fraction
          PPO for the transitional Copolymer D/PPO blends.  Identi-
          fication is the same as in Fig. 9.

That the compatible PPO blends appear to exhibit synergism in tensile strength can be interpreted in several ways. Extension of rubber network theory (27) would suggest that an increase in strength could result from an increase in the number of network chains per unit volume formed by the entanglements. Increased entanglement density may occur as a result of specific interchain interactions such as those reported in the case of PS/PPO blends by Wellinghoff et al (22). However, recent rheological studies of PS/PPO blends indicate that the dependence of blend entanglement molecular weight ($M_e$) on blend composition falls between the linear rule of mixtures given by Eq. 2 if E is replaced by $M_e$ and the series model in the similarly modified form of Eq. 3 (28). These results would suggest that in direct contrast to the experimentally found enhancement in tensile strength, the tensile strength of the blend should be at best a linear function of PPO content if the entanglement explanation is valid.

An alternate explanation for synergism in tensile strength may be taken from the same packing density argument proposed by Kleiner (2) to explain the observed synergism in modulus. Borrowing from the theory of the mechanics of crystal structure, Buchdahl (29) suggests that the (shear) strength of amorphous glassy polymers can be shown to be inversely proportional to the spacing between chain segments (i.e., packing density). In other words, the maximum in blend strength may be directly related to the maximum in blend density previously reported for these blends (6).

Alternately, it may be argued that the suppression of the β relaxation of PPO by PS as indicated by dynamic mechanical studies (3,4) raises the stress level required to activate significant strain softening. The importance of the β relaxation in controlling stress-activated processes has been revealed in the study of several poly(vinyl chloride) polymer blends (15,16).

In contrast to the compatible PPO blends, the transitional blends exhibit a sigmoidal dependence of σ on $V_{PPO}$ (Fig. 10). A minimum is reached between 10 and 20% PPO while at 60 to 80% PPO a maximum appears to exist at the same level and in the same composition range as found for the compatible blends. The embrittlement transition is again apparent at high $V_{PPO}$ but more samples fail in the brittle mode at between 60 and 80% PPO than at corresponding blend compositions in the case of the compatible blends.

The incompatible blends of PpClS/PPO, exhibit a much broader minimum in σ (Fig. 11) but again σ appears to reach a synergistic level at 80% PPO. More samples fail in the brittle mode at corresponding compositions than do the compatible and transitional blends in the embrittlement region. Additional data is required

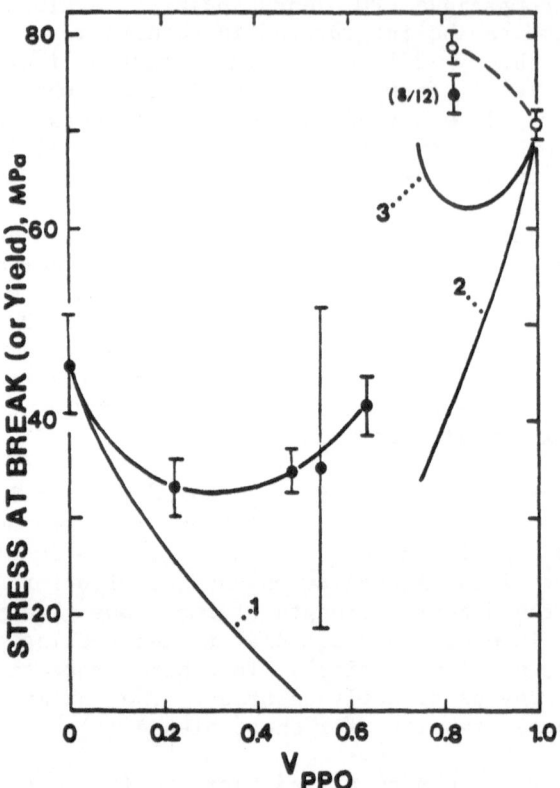

Fig. 11   Tensile stress at break (or yield) versus volume fraction
          PPO for the incompatible PpClS/PPO blends.   Identification
          is the same as in Fig. 9.   Curve 1, the Schrager model for
          poor adhesion (Eq. 6, r = 2.66); curve 2, data for untreat-
          ed glass bead/PPO composites (Ref. 24); curve 3, silane
          treated glass bead/PPO composites (Ref. 24).

in the 80–100% PPO region to ascertain whether in fact a local minimum in σ may occur in the yield region as is clearly observed in
the brittle zone or if in fact the high σ at 80% PPO is a true
synergistic effect.   Measurement of compressive strengths over a
wider range of blend compositions is presently in progress and
should provide a more conclusive picture.

    If the modulus behavior suggests as previously illustrated
that the incompatible PpClS/PPO blends behave as composites, then

the tensile strength of these blends may be expected to follow classical composite behavior as well. As discussed below, the dependence of $\sigma$ on $V_{PPO}$ as shown for PpClS/PPO in Fig. 11 appears more complicated than expected for a simple composite with controlled dispersed phase composition, size, and aggregation as may be achieved for PPO/glass beads composites as an example.

In modeling filled polymer composites, Schrager (30) has recently proposed an equation useful for predicting the break strength in the case where the filler is not treated to improve adhesion. The form of the Schrager equation is given as

$$\sigma = \sigma_o \, \exp(-rV) \qquad\qquad\qquad (6)$$

where $\sigma_o$ is the strength of the matrix polymer, V is the volume fraction filler, and r is an interfacial parameter which is found to be 2.66 for many composites including PPO/glass beads.

As illustrated in Fig. 11, the $\sigma$ of the PpClS/PPO blends agree with values predicted by Eq. 6 only at low loadings, i.e. $V_{PPO}<0.05$. At higher $V_{PPO}$, the observed blend $\sigma$ is much larger than values predicted by Eq. 6. If one assumes that the interfacial adhesion between the PpClS and PPO phase is greater that could be expected in the case of a polymer matrix and an untreated inorganic filler for which Eq. 6 is valid, then the inadequacy of Eq. 6 to model the PpClS/PPO blend results is not surprising.

In the case of perfect adhesion as for example when silane coupling agents are used to bond the filler to the matrix, Nielsen (31) has suggested that the elongation to break of the composite ($\varepsilon$) may be approximated by the following simple equation relating the elongation to break of the matrix ($\varepsilon_o$) and the volume fraction filler.

$$\varepsilon \approx \varepsilon_o (1-V)^{1/3} \qquad\qquad\qquad (7)$$

Composite break strength is then calculated by substituting Eq. 7 and a composite model equation for modulus into a linear stress-strain relation ($\sigma = E\varepsilon$). Qualitatively, the resulting expression predicts that $\sigma$ should rapidly fall to a minimum at ca. 10% filler and then increase to values equal to or greater than $\sigma_o$ depending upon which expression for modulus is substituted into the Hookean relation.

Quantitatively, the Nielsen expression for elongation to break (Eq. 7) has been reported to give good fit for some composites (32) but underestimates $\varepsilon$ in others such as the silane treated glass beads/epoxy system studied by Kenyon and Duffy (33).

In addition, Piggott and Leidner (34) have criticized the validity
of the simple geometrical considerations upon which Eq. 7 is based.
As shown in Fig. 12 for the composition range in which brittle
failure is observed to occur in the PpClS/PPO blends, experimental
values for break elongation consistently fall far above the curve
calculated by use of Eq. 7.  The usefulness of Eq. 7 in predicting

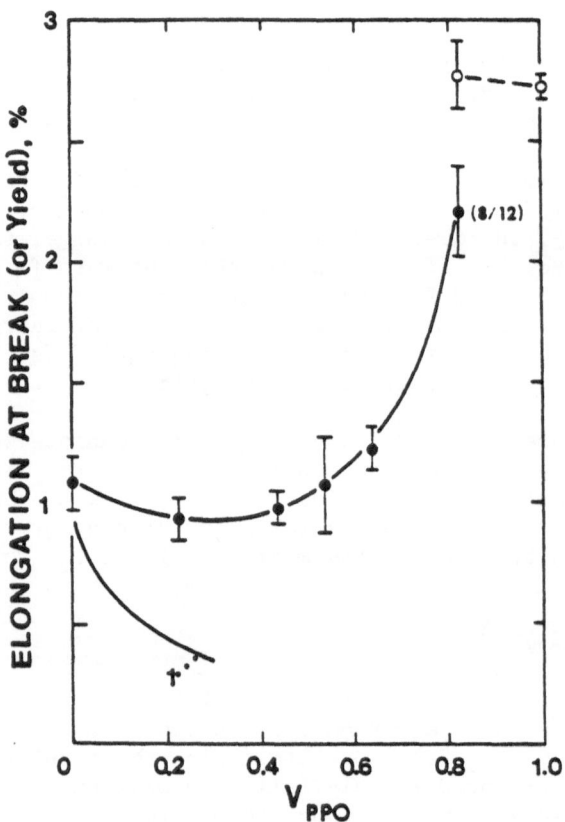

Fig. 12   Percent elongation at break (yield) for the incompatible
          PpClS/PPO blends.  ( ● ), elongation at break; ( ○ ),
          elongation at yield.  Values within parentheses indicate
          the fraction of samples failing in the principal mode in
          the embrittlement region.  Error bars indicate 95% con-
          fidence intervals.  Curve 1 was drawn from values calcu-
          lated from the Nielsen model for perfect adhesion compo-
          sites (Eq. 7).

$\varepsilon$ (and therefore $\sigma$) in these incompatible polymer blends is inherently restricted by the basic assumption upon which Eq. 7 is derived - i.e., that the dispersed phase is infinitely rigid. In the case of glass filled composites, this assumption can be considered adequate but in the case of polyblends for which the ratio of the moduli of the dispersed phase and the matrix polymer is nearly unity (a ratio of 1.3 in the PpClS/PPO blends), only a qualitative representation of break elongation  and therefore break strength can be expected. In consideration of these limitations, it is noted that the minimum in break strength that occurs at ca. 15% PPO in the transitional Copolymer D/PPO blends (Fig. 10) is in qualitative agreement with the Nielsen composite model for perfect adhesion between dispersed phase and matrix. By comparison, the minimum in break strength for the case of the incompatible PpClS/ PPO blends (Fig. 11) is located at higher $V_{PPO}$ and is deeper suggesting the probable importance of blend compatibility in determining the strength of the interfacial adhesion and thereby the break strength of the polyblend composite.

Additional differences in the compositional dependence of the tensile strength between incompatible polymer blends and polymer composites are revealed by directly comparing tensile strengths of PpClS/PPO blends ($V_{PPO} \geq 0.8$) with those of PPO filled with glass beads (1-30 μm) as given by Wambach et al.(24)  Values for E and $\sigma_y$ of unfilled PPO (2.55 GPa and 76.5 MPa, respectively) agree within confidence limits with those given in Table I at the same nominal strain rate and at nearly equivalent temperatures. In the upper right-hand portion of Fig. 11, the experimental yield strength of these glass bead/PPO composites is plotted versus $V_{PPO}$ ($=1-V_{beads}$) for the case in which the beads are untreated (curve 2) and the break strength when a silane coupling agent is used (curve 3). In the former case, the curve closely agrees with the model proposed by Schrager (Eq. 6) for poor adhesion while the latter curve (qualitatively) agrees with the Nielsen prediction for a minimum at 15% filler for the case of perfect adhesion (silane coupling) between the dispersed phase and matrix. In striking contrast are the corresponding values for stress at yield for PPO with 20% PpClS "filler". Instead of the expected reduction in yield strength due to the presence of a filler, the blend strength is actually about 11% higher than $\sigma_y$ of unfilled PPO and is comparable to the values found for the compatible and transitional blends at the same blend compositions.

In addition, there are differences in the shape of the stress-strain curves between those of the glass bead/PPO composites and the PpClS/PPO incompatible blends. In the case of glass bead/PPO, (32) and other glass bead filled polymer composites, (35,36) a knee or discontinuity has been observed at a stress corresponding to 23 MPa ($\varepsilon$= 0.35%) (32), independent of

filler concentration. This threshold stress corresponds to an
onset of stress whitening due to crazing and only occurs when the
beads are untreated (i.e., no coupling agent). These untreated
beads have poor adhesion with the PPO matrix and being unable to
support the tensile load act as stress concentrators. As evident
from the stress-strain curves for the PpClS/PPO blends (Fig. 4),
no such discontinuity is observed for the polymer blends. This
further supports the evidence from the behavior of the break
strength of these blends that there is strong interfacial adhesion
between phases in the incompatible PPO blends.


CONCLUSIONS

1.  Both compatible and semi-compatible PPO blends that show blend
    densification exhibit a small synergistic maximum in their
    modulus-composition plots which can be modeled by the classical
    rule of mixtures for composites with an additional interaction
    term. The incompatible PPO blends exhibit no blend densifica-
    tion and can be modeled adequately by the series model for com-
    posites.

2.  In terms of the relative magnitude of the size of the effect,
    tensile stress and elongation at break (or yield) are more
    sensitive than modulus to changes in the compatibility of the
    blend. The compatible PPO blends exhibit a maximum in strength
    at high PPO blend compositions. The semi-compatible (transi-
    tional) and incompatible blends exhibit both a minimum in the
    blend composition range in which brittle fracture predominates
    and an apparent maximum in the ductile zone. Qualitative
    application of the Nielsen model for break strength of com-
    posites suggests that the minimum in break strength occuring
    at about 15% PPO in the semi-compatible PPO blends is indica-
    tive of strong interfacial adhesion between phases.

3.  Although some insight into the mechanical properties of incom-
    patible polymer blends can be gleaned from an understanding
    of the way polymer composites with rigid fillers behave, the
    full picture is more complex. Factors that influence mechani-
    cal properties such as the shape, size, degree of agglomera-
    tion, and interfacial adhesion of the dispersed phase (37)
    intimately depend upon particular processing conditions and
    the degree of miscibility of components in the blend. An
    additional complication in the particular case of incompatible
    polymer blends is the occurence of phase inversion and the
    possibility of two continuous phases. Additional controlled
    studies may suggest ways that immiscible polymers may be
    blended to achieve attractive mechanical properties.

REFERENCES

1. D.R. Paul, C.E. Vinson, and C.E. Locke, Polym. Eng. Sci., 12, 157 (1972)
2. a) L.W. Kleiner, F.E. Karasz, and W.J. MacKnight, SPE 36th Annual Technical Conference, Washington, D.C., April 24, 1978, pp. 243-248; Polym. Eng. Sci., 19 (7), 519 (1979); b) L.W. Kleiner, Ph.D. Dissertation, University of Massachusetts, 1976
3. A.F. Yee, Polym. Prepr., Amer. Chem. Soc., Div. Polym. Chem., 17 (1), 145 (1976); Polym. Eng. Sci., 17, 213 (1977)
4. S. Wellinghoff and E. Baer, Amer. Chem. Soc., Div. Org. Coatings Plast. Chem., Pap. 36 (1), 140 (1976); J. Appl. Polym. Sci., 22, 2025 (1978)
5. W.J. MacKnight, F.E. Karasz, and J.R. Fried, in "Polymer Blends" (D.R. Paul and S. Newman, eds.), Vol. I., pp. 185-242, Academic Press, New York, 1978
6. J.R. Fried, F.E. Karasz, and W.J. MacKnight, Macromolecules, 11, 150 (1978)
7. a) F.E. Karasz, W.J. MacKnight, and J.J. Tkacik, Polym. Prepr., Amer. Chem. Soc., Div. Polym. Chem. 15 (1), 415 (1974); b) J.J. Tkacik, Ph.D. Dissertation, University of Massachusetts, 1976
8. A.R. Shultz and B.M. Beach, Macromolecules, 7, 902 (1974)
9. P.R. Couchman, Macromolecules 11 (6), 1156 (1978)
10. R.E. Wetton, W.J. MacKnight, J.R. Fried, and F.E. Karasz, Macromolecules, 11 158 (1978 )
11. J. Letz, J. Polym. Sci., Polym. Phys. Ed., 7, 1987 (1969)
12. B. Maxwell, SPE J., 28 (2), 24 (1972)
13. J.R. Fried, Ph.D. Dissertation, University of Massachusetts, 1976
14. W.J. Jackson, Jr., and J.R. Caldwell, J. Appl. Polym. Sci., 11, 227 (1967)
15. N. Sundgren, G. Bergman, and Y.J. Shur, J. Appl. Polym. Sci., 22, 1255 (1978)
16. G. Bergman, H. Bertilsson, and Y.J. Shur, J. Appl. Polym. Sci., 21, 2953 (1977)
17. E.P. Cizek, U.S. Patent 3,383,435 (assigned to General Electric) August 11, 1967
18. L.E. Nielsen, "Mechanical Properties of Polymers and Composites" Marcel Dekker, Inc., New York, 1974, p. 397
19. Ibid., p. 401
20. L.J. Cohen and O. Ishai, J. Compos. Mater., 1, 390 (1967)
21. L.E. Nielsen, "Predicting the Properties of Mixtures: Mixture Rules in Science and Engineering", Marcel Dekker, Inc., New York, 1978, p. 22
22. S.T. Wellinghoff, J.L. Koenig, and E. Baer, J. Polym. Sci., Polym. Phys. Ed., 15, 1913 (1977)
23. K.L. Trachte and A.T. DiBenedetto, Int. J. Polym. Mater., 1, 75 (1971)

24.  A. Wambach, K. Trachte, and A. DiBenedetto, J. Comp. Mater.,
     2, 266 (1968)
25.  H. Van Oene, in "Polymer Blends" (D.R. Paul and S. Newman, eds.)
     Vol. 1, pp. 295-352, Academic Press, New York, 1978
26.  L. Nicolais and A.T. DiBenedetto, J. Appl. Polym. Sci., 15,
     1585 (1971)
27.  F. Bueche, "Physical Properties of Polymers", New York,
     Wiley-Interscience, 1962, p. 202
28.  L.R. Schmidt, J. Appl. Polym. Sci., 23, 2463 (1979)
29.  R. Buchdahl, J. Polym. Sci., 28, 239 (1958)
30.  M. Schrager, J. Appl. Polym. Sci., 22, 2379 (1978)
31.  L.E. Nielsen, J. Appl. Polym. Sci., 10, 97 (1966)
32.  R.E. Lavengood, L. Nicolais, and M. Narkis, J. Appl. Polym.
     Sci., 17, 1173 (1973)
33.  A.S. Kenyon and H.J. Duffy, Polym. Eng. Sci., 7, 189 (1967)
34.  M.R. Piggott and J. Leidner, J. Appl. Polym. Sci., 18, 1619
     (1974)
35.  L. Nicolais and M. Narkis, Polym. Eng. Sci., 11 (3), 194
     (1971)
36.  L. Nicolais, E. Drioli, and R.F. Landel, Polymer, 14, 21
     (1973)
37.  G.W. Brasell and K.B. Wischmann, SPE Regional Technical
     Conference on Advances in Reinforced Thermoplastics, El
     Segundo, California, Paper 1, pp. 1-16 (1972)

# CRYSTALLIZATION FROM MISCIBLE POLYMER BLENDS

D. R. Paul and J. W. Barlow

Department of Chemical Engineering
The University of Texas at Austin
Austin, TX    78712

## INTRODUCTION

Until a few years ago, most of the recognized examples of miscible polymer blend pairs involved only amorphous components. However, recently numerous blend systems have been identified in which one or both components are crystallizable. The systems of interest here are miscible in the melt state, but upon cooling one or more of the components separates from the mixture as a pure crystalline phase. This form of solid-liquid phase separation represents a different situation than a liquid-liquid miscibility gap since complete miscibility may still exist in the remaining amorphous phase. The objective of this paper is to review some of the pertinent fundamental issues and recent results for miscible blends where crystallization is possible.

Such systems are of interest for several reasons. First, blending may offer a useful way to control the rate of crystallization and, consequently, the level of crystallinity developed under particular processing circumstances. Second, crystallization of one of the components from a miscible blend may create certain product problems. An example of this is the unwanted hardening of PVC compounds when a polymeric plasticizer crystallizes (1-4). Finally, observation of crystallization may provide a useful way to obtain fundamental information about a miscible blend.

Co-crystallization of components of a metallic alloy is well known; however, there seem to be no clearly established examples of this in polymer blend systems (5,6). In the discussions that follow, it is assumed that the crystalline phases which develop are not mixtures.

THERMODYNAMICS

It is instructive to examine the equilibrium thermodynamics of miscible polymer blend systems which contain a crystallizable component. When both components have molecular weights normal for commercial polymers, the configurational entropy of mixing is quite small. For present purposes, it will be an adequate approximation to regard the entropy of mixing as zero and to equate the free energy of mixing with the enthalpy of mixing.

$$\Delta G_{mix} \cong \Delta H_{mix} = B\phi_1\phi_2 \qquad\qquad 1$$

where in eq. 1 we have adopted a van Laar expression to represent the composition dependence ($\phi_i$ = volume fraction of i) of the heat of mixing. The parameter B characterizes the magnitude of the heat of mixing and will be negative for miscible blends. Recent results (7-16) have shown that B may be as large as -4 cal/cm$^3$ but more typically might be smaller in magnitude than this.

The free energy of crystallization for a pure polymer may be approximated as follows (17):

$$\Delta G_{crys} = - H_f^o \left(1 - \frac{T}{T_m}\right)\phi_c \qquad\qquad 2$$

where $T_m$ = melting point, $\Delta H_f^o$ = enthalpy of fusion for 100% crystalline polymer, and $\phi_c$ is the fractional crystallinity developed and is a volume fraction in the units employed here. For many polymers, $\phi_c$ is of the order of 0.5 while $\Delta H_f^o$ typically falls within the range of 20 to 60 cal/cm$^3$ (17).

Fig. 1 gives a graphical comparison of the magnitudes of these free energy changes for a typical case. Curve A establishes pure amorphous components 1 and 2 as the reference state. Curve B shows the heat of mixing computed from eq. 1 with B = -3 cal/cm$^3$ -- a relatively strong energetic interaction between components. The maximum $\Delta H_{mix}$ of -0.75 cal/cm$^3$ occurs at $\phi_1 = \phi_2 = 0.5$. Line C shows the reduction in free energy from the unmixed reference state caused by crystallization of component 2. Here, we have set $\Delta H_f^o$ = 40 cal/cm$^3$, $\phi_c$ = 0.5, a temperature such that $(T_m - T)/T_m$ = 0.25. The latter would lead to a near maximum rate of crystallization for most polymers and corresponds to T of about 40°C for a polymer with a melting point of 150°C.

Now it is of interest to examine the free energy of a mixture of polymers with these mixing and melting characteristics. Since there is no suitable theory available to predict equilibrium crystallinity, we have done the following. First, we have assumed $\phi_c$ = 0.5 for pure 2 as mentioned earlier. Next, we have assumed

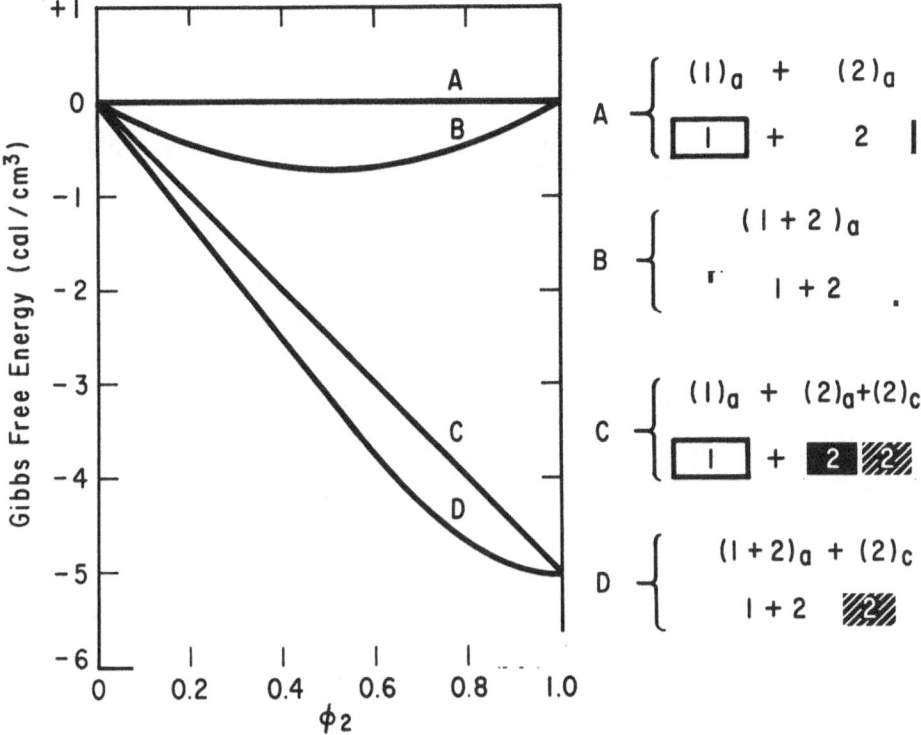

Fig. 1.  Free energy for miscible blends with and without crystal-
         linity of component 2 using parameters described in text.
         Notation on right schematically defines state considered:
         subscript c and shading denotes crystalline phase while
         subscript a and lack of shading denote amorphous  phase.

that the same fraction of 2 crystallizes for all blend compositions
-- a commonly observed situation as will be discussed later.
Because of this, the volume fraction of 2 in the remaining amorphous phase is

$$\phi_2^{(a)} = \frac{(1 - \phi_c)\phi_2}{\phi_1 + (1 - \phi_c)\phi_2} \qquad\qquad 3$$

rather than $\phi_2$ which corresponds to the overall blend composition
and it is the former which must now be employed with eq. 1 to
compute the heat of mixing in this case.  The free energy relative
to the reference state becomes

$$\Delta G_{mix} + \Delta G_{crys} = \frac{B(1 - \phi_c)(1 - \phi_2)\phi_2}{[1 - \phi_c\phi_2]^2} - \Delta H^\circ_f \phi_c \phi_2 \left(1 - \frac{T}{T_m}\right) \qquad 4$$

and is shown in Fig. 1 as Curve D. Thus, we see in this typical case that crystallization provides a means to obtain an even lower free energy than the mixed amorphous state and, hence, crystals of pure 2 will exist at equilibrium.

However, the above situation may not always be the case since, in principle, Curve B may lie <u>below</u> Curve D over a portion of the composition range for certain values of the parameters in eq. 4. In this situation, no crystals of component 2 will exist at equilibrium. This will be true when

$$\frac{(1 - \phi_2)(1 - 2\phi_2 + \phi_c\phi_2^2)}{[1 - \phi_c\phi_2]^2} > \frac{\Delta H^\circ_f}{(-B)} \left(1 - \frac{T}{T_m}\right) \qquad 5$$

which is possible when

$$1 > \frac{\Delta H^\circ_f}{(-B)} \left(1 - \frac{T}{T_m}\right) \qquad 6$$

Clearly, eq. 6 can be satisfied for any values of $\Delta H^\circ_f$ and $(-B)$ provided T is close enough to $T_m$. However, the values of $\Delta H^\circ_f$ for polymers are generally much larger than the $(-B)$ values we can expect for even quite strong specific interactions and, consequently, for T significantly below $T_m$, eq. 6 cannot be satisfied. Thus, we may expect for nearly all polymer blends that thermodynamics will not preclude crystallization of one of its components. However, as we will see later, kinetics may do this.

For temperatures near the melting point a different thermodynamic analysis becomes useful for characterizing the interaction between blend components (12,14). Because of the lower free energy in the mixed amorphous phase, the crystals of component 2 have a lower equilibrium melting point than pure component 2. The melting point depression stemming from the enthalpic part of the mixing process is

$$(\Delta T_m)_H = -T_{m2} \left(\frac{V_{2u}}{\Delta H_{2u}}\right) B \phi_1^2 \qquad 7$$

where

$$B = RT(\frac{X_1}{\tilde{V}_1})$$

8

Entropic contributions give an additional depression of

$$(\Delta T_m)_S = -R(\frac{V_{2u}}{\Delta H_{2u}})(T_{m2})^2\left[\frac{\rho_2 \ln \phi_2}{M_2} + (\frac{\rho_2}{M_2} - \frac{\rho_1}{M_1})\phi_1\right]$$

9

but this term is less than 1°C for component molecular weights larger than $10^4$. From melting point depression measurements, estimates of the interaction parameter B can be obtained as reported recently (12).

This simple analysis does not acknowledge the fact that the crystals may be finite in size and that the contribution of this effect may be variable with blend composition. One means of accounting for this issue is by use of the Hoffman-Weeks type analysis (14,16,18).

KINETICS

The crystallization rate process is governed by the degree of undercooling or driving force and the resistance to segmental motions. Because of these factors, crystallization at finite rates is limited to the region between $T_m$ and $T_g$. Blending two polymers can have important effects on this temperature gap $(T_m - T_g)$ available for crystallization as shown schematically in Fig. 2 (12,19). Here, we see on the left a case where $T_g$ increases upon adding the amorphous diluent. This will retard molecular motions, decrease the gap $(T_m - T_g)$, and, in general, tend to restrict crystallization. On the right is the case where $T_g$ decreases upon adding the amorphous diluent and, hence, the $(T_m - T_g)$ gap is broadened, molecular motion is easier, and crystallization may become more rapid.

Direct observation of spherulite growth rates during isothermal crystallization yields fundamental information about the kinetics of this process because the results can be compared with pertinent theories and basic parameters can be quantified. However, such experiments are tedious and only a few studies on blends have been reported. Of related interest are a number of reports on the crystallization of blends of fractions of the same polymer but of differing molecular weights. This approach is useful for assessing the role of overall chain mobility and has shown in some instances that molecular weight segregation occurs on crystallization (20). Of more direct interest are several studies in which blends of the same vinyl polymers differing in tacticity have been employed.

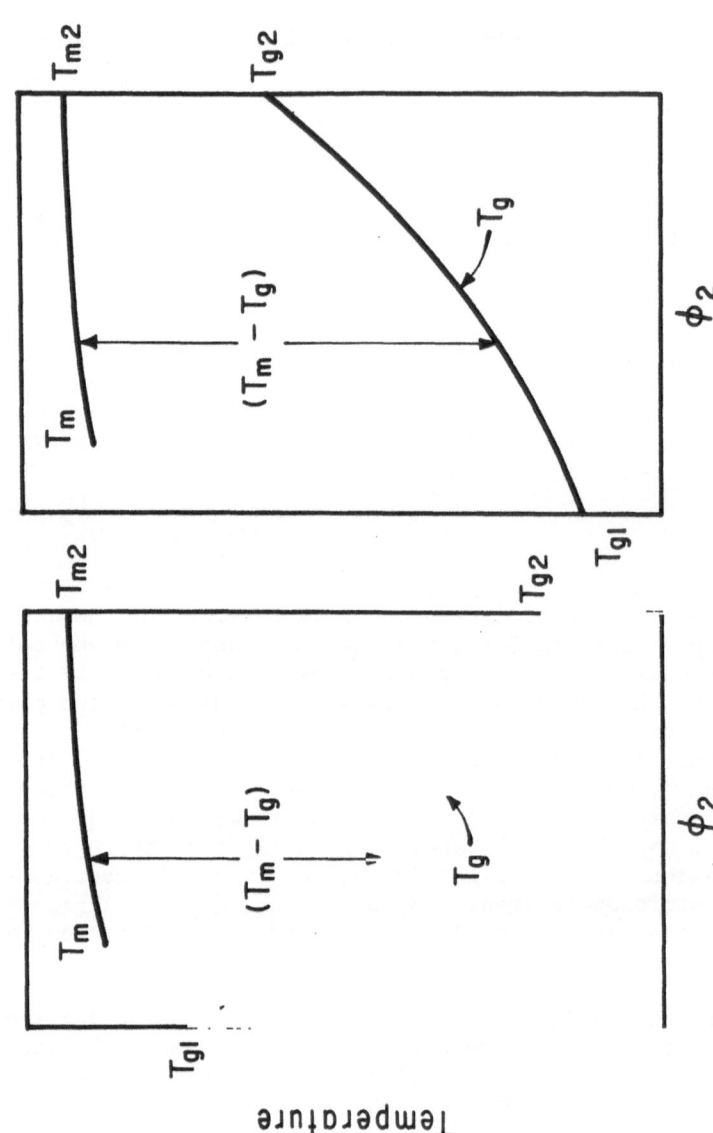

Fig. 2.   Possible relationships between $T_g$ and $T_m$ with blend composition for miscible systems.

Keith and Padden (21) reported the first studies in this area using atactic and isotactic polypropylene and atactic and isotactic polystyrene. More recently, others (22-24) have reported on the latter system. All agree that dilution by noncrystallizable, atactic polystyrene reduces the spherulite growth rate for the stereoregular isomer although there is some disagreement in detail. In general, it is felt that segregation of the isomers which must occur during crystallization and the associated transport processes are responsible for this effect. The isomers of polystyrene form a degenerate case of those situations depicted in Fig. 2 since $T_g$ is essentially independent of blend composition. However, it is interesting to wonder whether the isomers of polystyrene are in fact miscible in the melt or amorphous state. This point would be difficult to prove by most techniques; however, neutron scattering would seem to provide a viable approach.

Several fundamental studies on miscible blends of polymers with different chemical composition have been reported recently. Very limited data on the crystallization kinetics of isotactic polystyrene from mixtures with poly(phenylene oxide) have been reported by Berghmans and Overbergh (25). The most thorough study of this kind to date is that by Wang and Nishi (26) on crystallization of poly(vinylidene fluoride), $PVF_2$, from mixtures with poly(methyl methacrylate), PMMA. They found that addition of PMMA greatly reduced the rate of $PVF_2$ crystallization with the bulk of this effect stemming from the increased $T_g$ of the mixture (see left hand side of Fig. 2). Stein et al. (27) have presented a preliminary report on the system poly(ethylene terephthalate), PET, and poly(butylene terephthalate), PBT, which apparently exhibits miscibility. Both components can crystallize and do so separately from mixtures. Robeson (2) followed the crystallization of poly($\varepsilon$-caprolactone), PCL, from mixtures with poly(vinyl chloride), PVC, by observing the change in modulus of the blend. PCL crystallization rate greatly decreased with PVC content since this increased the $T_g$ of the mixture.

It is also of interest to consider other observations on crystallization behavior from miscible or partly miscible blend systems which are now available. Some of these involve the observation of a crystallization exotherm during heating or cooling at a constant rate during thermal analysis. While such results do not yield the fundamental information of the experiments mentioned above, they are nevertheless quite informative. A useful parameter of such observations is the temperature, $T_c$, at which the exotherm peaks. For rapidly crystallizing systems, this process is monitored during cooling. Addition of a miscible diluent has been shown to shift $T_c$ to lower temperatures (16) which is consistent with a reduction in crystallization rate. In some cases, crystallization is too slow and the glass transition is reached before any appreciable

crystallinity is developed. In these cases, it is common to observe
the exotherm of crystallization on heating. For several systems,
we have found that $T_c$ increased on dilution with a miscible diluent
(28,29). Such observations are especially interesting for partly
miscible systems. For example, we recently showed that PET is
completely miscible with polycarbonate in blends containing more
than about 70% PET although two amorphous phases always exist at
lower PET contents (28). Fig. 3 shows the response of $T_c$ after the
various thermal histories indicated. Interestingly, $T_c$ increases
on addition of PC up to the solubility limit as expected and decreases thereafter. On the other hand, PBT is partly miscible with
PC but not completely so at any composition (30). Heating of these
blends showed two separate crystallization exotherms which evidently
arise from PBT crystallization from two separate amorphous phases
after their respective glass transitions were surpassed.

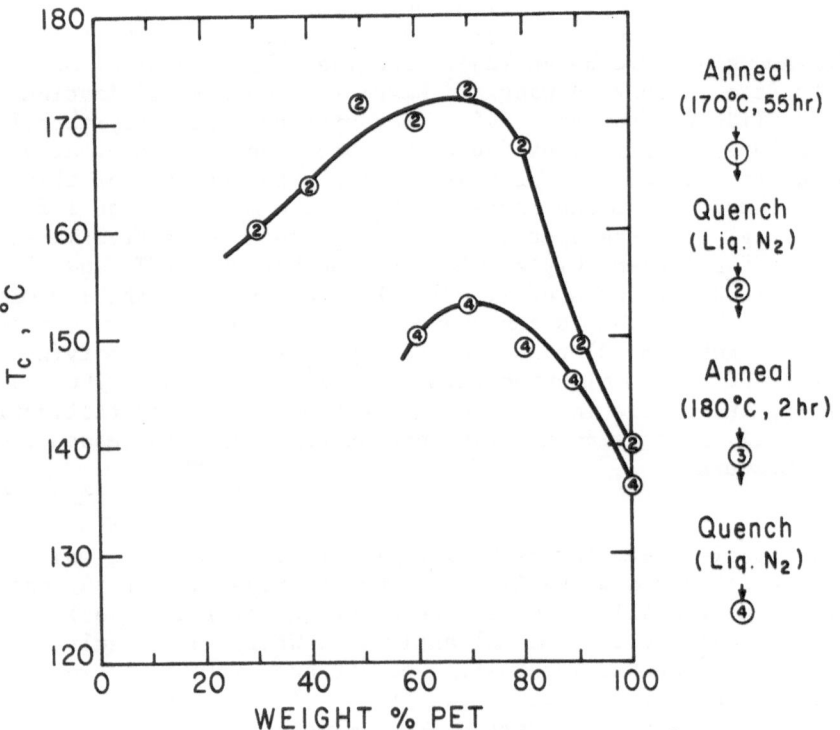

Fig. 3.   Variation of crystallization temperature on heating for
         PET-PC blends. Notation at right defines thermal history
         prior to heat i.

Another simple but useful approach for rapid assessment of
crystallization kinetics is cyclic thermal analysis.  In this technique, the sample is cyclicly cooled and heated at a constant rate
between temperature limits that lie beyond the $T_m$ and $T_g$.  Here, one
is interested in the magnitude of the endotherm as the sample melts
on heating which reflects the maximum crystallinity of component 2
developed in the cycle.  This procedure provides a complex but
finite time-temperature opportunity for the crystallization process,
and systems which crystallize slowly will not be able to develop the
full extent of crystallinity that might be possible during, say,
long term isothermal annealing at an appropriate crystallization
temperature.  Fig. 4 shows some of the patterns of behavior we have
observed for various miscible blend systems using the above scheme.
Here, the results are given as percent crystallinity of component 2
based on total blend mass.  Case (a) corresponds to the situation
where component 2 crystallizes to the same extent in the blend as
it does in the neat state.  This has been observed for some miscible
blends where the diluent did not appreciably retard crystallization
kinetics (9,10) and for some immiscible blend systems (11,12).

In case (b) in Fig. 4, some crystallinity is developed at all
blend compositions but the amount is somewhat reduced from that
in case (a).  Blends of a copolyester, KODAR, with PC have shown
this type of response (29).  In case (c), no crystallinity develops
during the particular time-temperature cycle for blends rich in the
miscible diluent.  Such behavior has been observed for $PVF_2$ blends
containing various miscible, oxygen containing polymers (12).
Case (d) is an interesting one in that some blends seem to develop
more crystallinity for component 2 than this polymer does alone.
This has been observed for blends of poly(1,4 cyclohexanedimethanol
terephthalate) with PC (29) and for $PVF_2$ blends with poly(vinyl
acetate) (10).  An even more intriguing situation is case (e), where
polymer 2 does not crystallize at all in the neat state but through
the plasticizing action of the miscible diluent, it does so in
blends.  This has been observed for PC, which does not crystallize
during normal melt processing, but does so when blended with poly-
($\varepsilon$-caprolactone) (31) and other low $T_g$ polyesters (13) as illustrated
in Fig. 5.  This situation can result when the transition behavior
follows the pattern shown on the right in Fig. 2.  Other systems
show minor variations of those patterns illustrated in Fig. 4 and
combinations of them when both components crystallize.

Clearly, the $T_g$ value for component 1 in miscible blends is
responsible for much of the crystallization behavior of component 2.
This is most clearly illustrated by our recent data for $PVF_2$ in
blends with a variety of miscible components whose $T_g$ varied rather
widely (12).  To illustrate this we return to case (c) in Fig. 4
where we define $(w_2)_0$ as the limiting composition at which crystallinity no longer develops during cyclic DTA.  Fig. 6 shows the

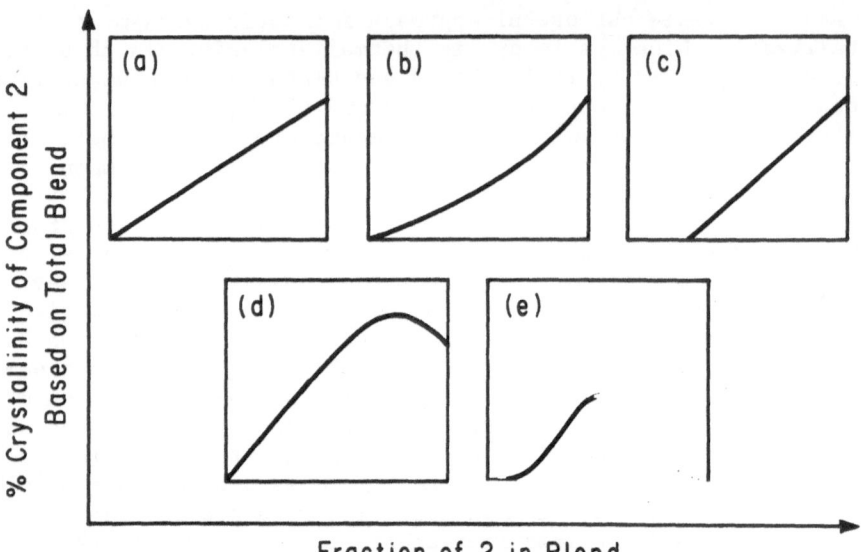

Fig. 4.    Classification of patterns of crystallinity development
during cyclic thermal analysis.  Only component 2 is
crystallizable in these cases.

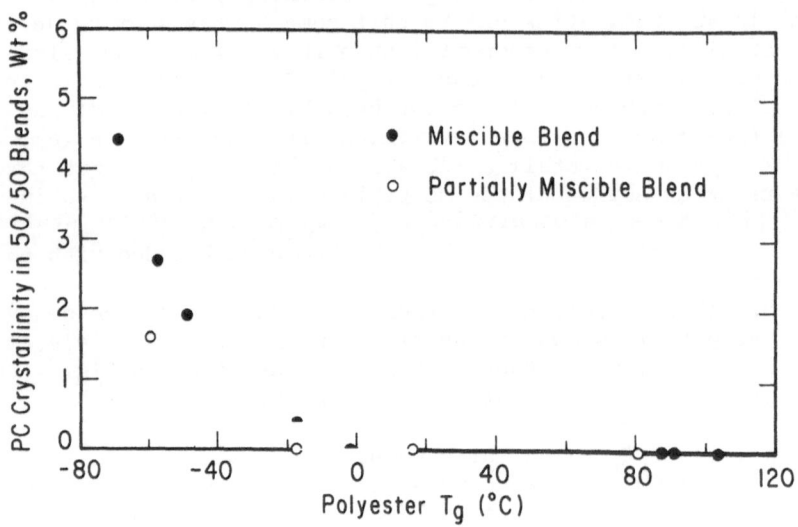

Fig. 5.    Crystallinity of polycarbonate developed in blends
with various polyesters.  PC crystallization occurs
when the polyester $T_g$ is low enough to cause
adequate plasticization.

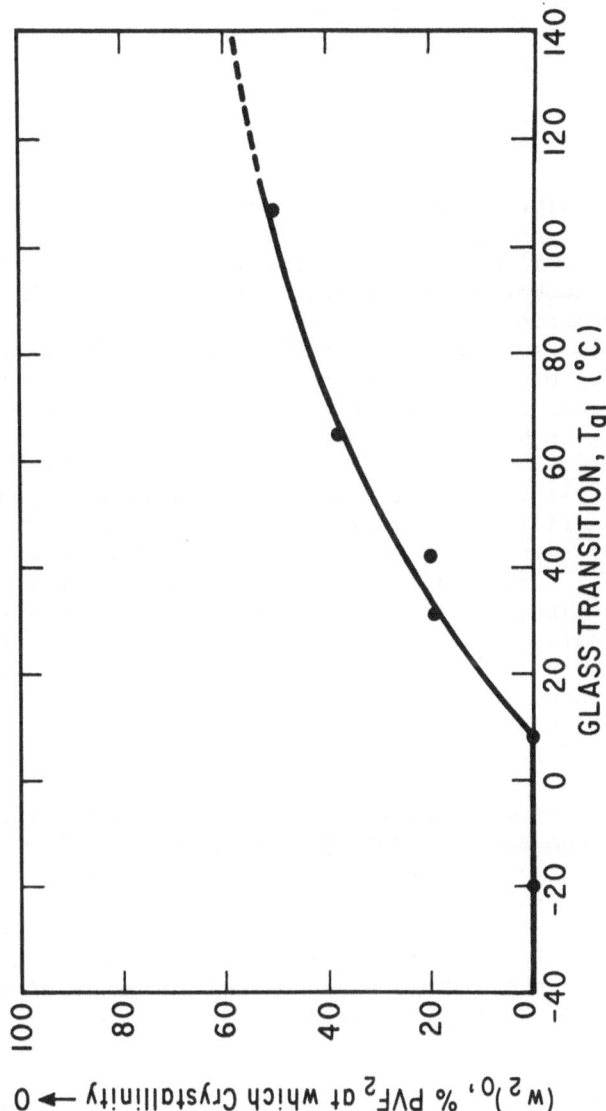

Fig. 6.   Effect of the $T_g$ of amorphous, miscible diluents on the limiting concentration of $PVF_2$ in blend below which no crystallinity develops during cyclic thermal analysis.

response of $(w_2)_0$ to the glass transition of the miscible diluent,
$T_{g1}$. When $T_{g1}$ is less than about 10°C, $(w_2)_0$ is always zero which
means case (c) has degenerated into case (a) of Fig. 4 and that the
$PVF_2$ is able to crystallize fully because the mobility of the blend
is no significant hindrance to its crystallization for the heating/
cooling rate of 10°C/min. However, when $T_{g1}$ is increased to higher
levels, $(w_2)_0$ takes on finite values. The higher $T_{g1}$, the smaller
is the $(T_m - T_g)$ gap and the more crystallization is retarded.


MORPHOLOGY

    Important morphological questions about miscible blends in
which one or both components have crystallized concern the nature
of the overall crystalline texture and the location of the amorphous material. A number of recent publications have sought
answers to these questions for particular systems. Early microscopy studies showed that noncrystallizable diluents increased the
coarseness of the spherulites of the crystallizable components
(21,22); however, more recent scattering experiments have been
able to determine more detailed information.

    Stein et al. (32) recently reported that PCL when blended with
PVC develops volume filling spherulites. With increasing PVC content, the amorphous layer thickness increased just enough to
accommodate the PVC. Hence, linear crystallinities found from
scattering measurements agreed well with bulk crystallinities
determined from density or DSC. Similar studies on blends of isotactic polystyrene with PPO suggest that the PPO is located in the
interlamellar amorphous regions of the polystyrene spherulites;
however, addition of the amorphous component causes some disordering
of the lamellae (33,34).

    Similar scattering observations on blends of atactic and iso-
tactic polystyrene suggest that segregation of the amorphous component occurs within the growing spherulites of isotactic polystyrene. Evidently, domains form which are larger than interlamellar
but smaller than spherulite size. Several authors (21,23) have
suggested that the scale of segregation, $\delta$, is related to the
relative kinetics of spherulite growth, G, and diffusion of the
noncrystallizable component, D, through

    $$\delta = D/G$$

In this kinetic view, the small scale interlamellar segregation seen
for i-PS/PPO and PCL/PVC blends is believed to be the result of the
higher $T_g$ of the amorphous component compared to the crystallizable
one which causes diffusion to dominate; whereas, in the i-PS/a-PS
system the decreased spherulite growth rate is believed to cause

segregation into regions of larger dimensions (23).  However, it would seem that the differences in the thermodynamics of these systems might also be a factor.

Stein et al. have described interesting morphologies for the PET/PBT system in which both components crystallize (27,35).  PET rich blends form PET spherulites containing PBT crystals while PBT rich blends develop PBT spherulites containing PET crystals. It will be interesting to learn more about this complicated but important system such as the location and nature of the amorphous material.

SUMMARY

Crystallization from blends is a quite different phase separation process than that in a liquid-liquid miscibility gap.  It seems quite unlikely that the energetic interaction between two polymeric components would ever be strong enough to preclude significantly the crystallization of one of the components of a miscible blend. However, kinetic factors may severely retard crystallization in some cases.  On the other hand, addition of a sufficiently mobile, miscible diluent may kinetically facilitate the crystallization of sluggish polymers like polycarbonate.  Apparently, component segregation on crystallization can occur on various size scales ranging from interlamellar to spherulite dimensions.

Interestingly, there seem to be no proven cases of co-crystallization of components from blends.  An important area for additional work is to determine the effect of non-miscible polymeric components on crystallization kinetics, ultimate crystallinity, and morphology.

ACKNOWLEDGEMENTS

Acknowledgement is made to the donors of the Petroleum Research Fund, administered by the American Chemical Society, for support of this research.

REFERENCES

1.  C. F. Hammer, Ch. 17 in "Polymer Blends, Vol. II," D. R. Paul
        and S. Newman, eds., Academic Press, New York, 1978.
2.  L. M. Robeson, J. Appl. Polym. Sci., 17, 3607 (1973).
3.  H. E. Bair, E. W. Anderson, G. E. Johnson, and T. K. Kwei,
        A.C.S. Polym. Preprints, 19(#1), 143 (1978).

4.  H. E. Bair, D. Williams, T. K. Kwei, and F. J. Padden, A.C.S.
    Organic Coatings and Plastics Chemistry Preprints, 37 (#1),
    240 (1977).
5.  D. R. Paul, Ch. 1 in "Polymer Blends, Vol. I," D. R. Paul and
    S. Newman, eds., Academic Press, New York, 1978.
6.  D. Krevor and P. J. Phillips, in "Polymer Alloys," D. Klempner
    and K. C. Frisch, eds., Plenum Press, New York, 1977.
7.  D. R. Paul and J. O. Altamirano, Adv. Chem. Ser., 142, 371
    (1975).
8.  R. L. Imken, D. R. Paul, and J. W. Barlow, Polym. Eng. Sci.,
    16, 593 (1976).
9.  D. C. Wahrmund, R. E. Bernstein, J. W. Barlow, and D. R. Paul,
    Polym. Eng. Sci., 18, 677 (1978).
10. R. E. Bernstein, D. R. Paul, and J. W. Barlow, Polym. Eng. Sci.,
    18, 683 (1978).
11. R. E. Bernstein, D. C. Wahrmund, J. W. Barlow, and D. R. Paul,
    Polym. Eng. Sci., 18, 1220 (1978).
12. D. R. Paul, J. W. Barlow, R. E. Bernstein, and D. C. Wahrmund,
    Polym. Eng. Sci., 18, 1225 (1978).
13. C. A. Cruz, Ph.D. Dissertation, The University of Texas at
    Austin, 1978.
14. T. Nishi and T. T. Wang, Macromolecules, 8, 909 (1975).
15. T. K. Kwei, G. D. Patterson, and T. T. Wang, Macromolecules,
    9, 780 (1976).
16. E. Roerdink and G. Challa, Polymer, 19, 173 (1978).
17. L. Mandelkern, "Crystallization of Polymers," McGraw-Hill,
    New York, 1964.
18. J. D. Hoffman and J. J. Weeks, J. Res. Nat'l. Bur. Stand.,
    Sec. A, 66, 13 (1962).
19. W. J. MacKnight, F. E. Karasz, and J. R. Fried, Ch. 5 in
    "Polymer Blends, Vol. I," D. R. Paul and S. Newman, eds.,
    Academic Press, New York, 1978.
20. J. H. Magill and H. M. Li, Polymer, 19, 416 (1978).
21. H. D. Keith and F. J. Padden, J. Appl. Phys., 35, 1270 and 1286
    (1964).
22. G. S. Y. Yeh and S. L. Lambert, J. Polym. Sci.: Part A-2, 10,
    1183 (1972).
23. F. P. Warner, W. J. MacKnight, and R. S. Stein, J. Polym. Sci.:
    Polym. Phys. Ed., 15, 2113 (1977).
24. R. S. Stein, F. P. Warner, A. Escala, E. Balizer, T. Russell,
    and J. Koberstein, A.C.S. Organic Coatings and Plastics
    Chemistry Preprints, 37(#1), 7 (1977).
25. H. Berghmans and N. Overbergh, J. Polym. Sci.: Polym. Phys.
    Ed., 15, 1757 (1977).
26. T. T. Wang and T. Nishi, Macromolecules, 10, 421 (1977).
27. A. Escala, E. Balizer, and R. S. Stein, A.C.S. Polym. Pre-
    prints, 19(#1), 153 (1978).
28. T. R. Nassar, D. R. Paul, and J. W. Barlow, J. Appl. Polym.
    Sci., 23, 85 (1979).

29.  R. N. Mohn, D. R. Paul, J. W. Barlow, and C. A. Cruz, J. Appl.
     Polym. Sci., 23, 575 (1979).
30.  D. C. Wahrmund, D. R. Paul, and J. W. Barlow, J. Appl. Polym.
     Sci., 22, 2155 (1978).
31.  C. A. Cruz, D. R. Paul, and J. W. Barlow, J. Appl. Polym. Sci.,
     23, 589 (1979).
32.  F. B. Khambatta, F. Warner, T. Russell, and R. S. Stein, J.
     Polym. Sci.: Polym. Phys. Ed., 14, 1391 (1976).
33.  W. Wenig, F. E. Karasz, and W. J. MacKnight, J. Appl. Phys.,
     46, 4194 (1975).
34.  R. Hammel, W. J. MacKnight, and F. E. Karasz, J. Appl. Phys.,
     46, 4199 (1975).
35.  A. Escala and R. S. Stein, Adv. Chem. Ser., 176, 455 (1979).

An Analysis of the Thermal Degradation Under Processing

Conditions of ABS/PVC Blends

W. I. Congdon[1], H. E. Bair[2], and S. K. Khanna[3]

Bell Telephone Laboratories
2525 Shadeland Ave. [1,3]
Indianapolis, Indiana 46206
600 Mountain Ave.
Murray Hill, New Jersey 07974[2]

## Introduction

Blends of Polyvinylchloride (PVC) and Acrylonitrile-Butadiene-Styrene (ABS) offer a combination of rigidity, impact resistance, solvent resistance and self extinguishing properties which give these mixtures great utility throughout the plastics industry.

During the processing of certain of these blends we have seen an anomalous decrease in thermal stability of the blends over either pure component. In this study we report an analysis of the changes which have occurred in the separate phases of these polymer mixtures during melt processing.

## Experimental

The following materials were used as supplied:

| | |
|---|---|
| Epon 828 | Shell Chemical |
| Cycolac T-1000 (ABS) | Borg-Warner |
| Loxiol G30 and G70 | Henkel Inc. |
| Geon 103EP (PVC) | B. F. Goodrich |
| Thermolite 133 | M & T Chemical |
| Santicizer 7, 11 | Monsanto |

The thermal analytical data was measured with a DuPont 990 using the DSC or TGA (Model 951) module and verified using a Perkin Elmer DSC-2. The analytical methods used have been previously reported.[1]

Infrated data using the methods of A. P. Weir et al.[2],

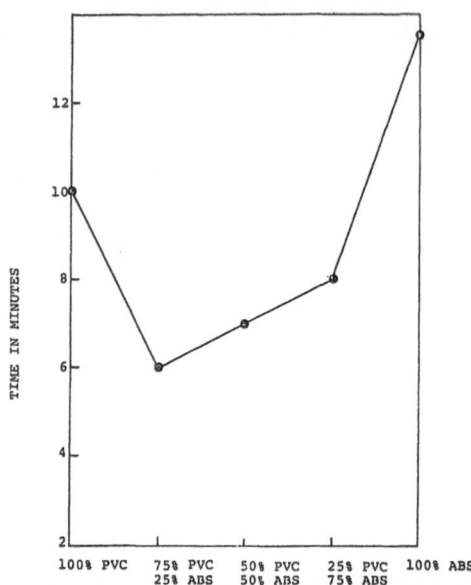

Figure 1.  A Plot of Time to Decomposition Versus Composition
For Blends of ABS/PVC In a Torque Rheometer, 70 RPM,
230° C.

was measured in transmission on the Perkin Elmer 621 grating
infrared spectrophotometer.  The C. W. Brabender Torque
Rheometer was used for process simulation.  In each experiment the temperature was pre-equilibrated, the rotors turning at 70 rpm and then a 45 gram sample charged in one slug.
This procedure essentially conforms to standard practice on
this rheometer.  250 mg samples were removed from the chamber
at frequent intervals for analysis.

## Blend Preparation

The following procedure was used to prepare all blends.
In a commercial Waring Blender, 93.5 grams of Geon 103EP
was charged along with 1 gram of Loxiol G70 and 0.5 grams of
Loxiol G30, two fatty acid ester type lubrican.  This mixture
was blended at high speed for five minutes and then the
following liquids were added:

| | |
|---|---|
| 2.5 grams | Thermolite 133, a Tin stabilizer from M & T |
| 0.5 grams | Epon 828, a Shell epoxy stabilizer |
| 2.0 grams | Santicizer 7,11, a Monsanto phthalate plasticizer, processing aid |

TABLE I

Thermal Analysis of the Processing of Cycolac
T1000 at 230°C in a Torque Rheometer

| Processing Time (Minutes) | PBD Tg | $\dfrac{Cal}{Cp} \cdot {}^{o}C\text{-gm}$ | SAN Tg | $\dfrac{Cal}{CP} \cdot {}^{o}C\text{-gm}$ |
|---|---|---|---|---|
| 0 | -84°C | 0.017 | 106°C | 0.064 |
| 7 | -84°C | 0.015 | 108°C | 0.066 |
| 10 | -85°C | 0.017 | 109°C | 0.069 |
| 15 | -82°C | 0.017 | 110°C | 0.065 |
| 20 | -83°C | 0.017 | 109°C | 0.069 |
| 25 | -82°C | 0.016 | 109°C | 0.072 |
| 30 | -81°C | 0.017 | 109°C | 0.069 |

$$\overline{X} = 0.0166 \, \frac{Cal}{{}^{o}C\text{-gm}} \qquad \overline{X} = 0.0677$$

$$= 7.9 \times 10^{-4} \qquad\qquad = 0.0028$$

$$4.7\% \qquad\qquad\qquad 4.2\%$$

Then high speed blending was resumed for another five
minutes. The resulting mixture was then weighed and fluxed
on a two roll mill with the appropriate amount of Cycolac
T-1000. The mill was heated to 350°F and the blend was given
three passes (five minutes total) then sheeted and ground
into course powder in a Thomas Wiley Laboratory mill model
4.

Results and Discussion

Both PVC and ABS degradation is known to be accompanied
by crosslinking which has been shown to cause the resin melt
viscosity to increase markedly. [3,4] The length of time an
individual compound is processed at elevated temperatures in a
torque rheometer before an increase in viscosity is detected
is a measure of the sample's thermal stability. In Figure 1,
the time until an increase in viscosity was detected on a
torque rheometer is plotted against PVC/ABS blend composition.
All three blends exhibited lower stability than either pure
component. Of the three blends the shortest time (6 minutes)
occurred for the 75% PVC/25% ABS blend while the longest
(8 minutes) time was registered for the 25% PVC75% ABS.

The unexpected reduction in stabilization, shown as a minimum
in Figure 1, could result from several causes. One hypothesis
would attribute the effect to local shear heating effects which
are induced by the higher viscosity ABS in the blends and could
accelerate dehydrohalogenation and crosslinking of the PVC
phase. Alternately, one could envision a synergistic chemical
interaction between PVC and ABS phases which might lead to

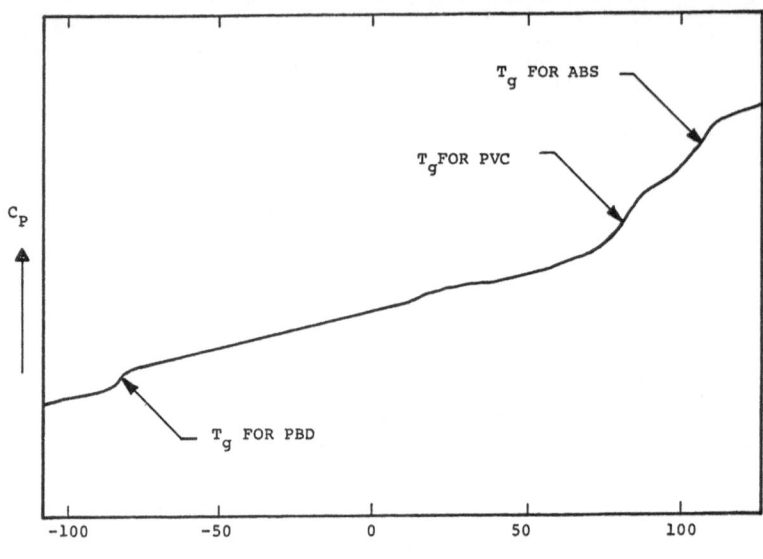

TEMPERATURE $^{\circ}$C

Figure 2.   A Plot of Heat Capacity ($C_p$) As A Function of Temperature
            For A 50/50 Mixture of ABS/PVC

TABLE II

Thermal Analysis of the Processing of a 50/50 Blend
of ABS/PVC in a Torque Rheometer at 230$^{\circ}$C

| Processing Time Minutes | PBD Tg | Cp $\frac{Cal}{^\circ C\text{-gm}}$ | PVC Tg | Cp $\frac{Cal}{^\circ C\text{-gm}}$ | San Tg | Cp $\frac{Cal}{^\circ C\text{-gm}}$ |
|---|---|---|---|---|---|---|
| 0 | -87$^{\circ}$C | 0.0100 | 80$^{\circ}$C | 0.028 | 104 | 0.030 |
| 4.5 | -86$^{\circ}$C | 0.009 | 81$^{\circ}$C | 0.029 | 106 | 0.027 |
| 6.0 | -86$^{\circ}$C | 0.008 | 81$^{\circ}$C | 0.027 | 104 | 0.026 |
| 7.5 | -80$^{\circ}$C | 0.007 | 82$^{\circ}$C | 0.028 | 106 | 0.025 |
| 8.5 | -82$^{\circ}$C | 0.006 | 82$^{\circ}$C | 0.025 | 107 | 0.024 |
| 9.0 | -84$^{\circ}$C | 0.006 | 82$^{\circ}$C | 0.025 | 106 | 0.024 |
| 12.0 | -82$^{\circ}$C | 0.006 | 83$^{\circ}$C | 0.028 | 105 | 0.023 |
| 15.0 | -82$^{\circ}$C | 0.006 | 84$^{\circ}$C | 0.029 | 109 | 0.023 |
| 30.0 | -82$^{\circ}$C | 0.006 | 84$^{\circ}$C | 0.023 | 110 | 0.021 |
| 35.0 | -84$^{\circ}$C | 0.005 | 88$^{\circ}$C | 0.024 | 112 | 0.019 |

premature degradation of the ABS and crosslinking. The latter,
more complex hypothesis appears to be supported by the
observation that the blend with the most PVC (75%) is least
stable.

In order to probe this phenomenon various samples were
Brabended over a period of time at 230°C, 70 rpm and samples
removed periodically and examined via DSC. The ABS control
experiment (See Table 1) yielded the two expected Tg values
at -84°C and 104°C corresponding to the polybutadiene and
styrene-acrylonitrile phases of the ABS resin. Cp measurements for the unprocessed sample are in good agreement with
earlier values.[1] Subsequent Brabending of the control for up
to 30 minutes produced a 3°C increase in Tg and no detectable
lowering of Cp for either phase. If degradation had been
significant Bair et al. has shown in recent work[5] that ABS
oxidation, which occurred most readily in the PBD phase,
resulted in an increase and broadening in Tg as well as a
decrease in Cp. Thus we conclude that only small amounts
of degradation have occurred in either phase.

In Figure 2, a DSC curve for the 50% PVC/50% ABS blend
which has not been mixed in the Brabender is shown. Three
discontinuities in Cp were detected at -87°C, 80°C and 104°C
which correspond to PBD, PVC and SAN, respectively and
indicate that the PVC and ABS are immiscible. Thermal
analysis of the 50/50 blends after various periods of mixing
at 230°C were obtained and are shown in Table 2 and Figure 3.

After 9 minutes of mixing in the torque rheometer at
230°C the polybutadiene phase as measured by Cp at the
glass transition is reduced by 40%, the SAN phase by 20% and
the PVC phase by approximately 9%. Clearly the PBD phase is
degraded more rapidly under these conditions than either the
SAN or PVC phase. The SAN and PVC Cp at Tg for the blend
is more difficult to measure since the transitions partially
overlap. Nevertheless an unmistakable trend can be seen form
the data which shows that mixing of PVC with ABS at 230°C
leads to accelerated deterioration of both ABS phases but
most significantly to the rubber component of the blend.

In order to confirm our thermal analytical data we used
the infrated method of A. P. Weir et al.,[2] to quantitate
changes in composition of the blends after processing for 35
minutes at 230°C. Quadruplicate measurements in transmission at wavelengths of 755cm$^{-1}$ (Styrene), ]430cm$^{-1}$ (PVC),
1645cm$^{-1}$ (Butadiene) and 2220cm$^{-1}$ (Acrylonitrile) indicated
losses of 25+ 10% styrene, -10 ± 10% PVC, 55 ± 16% PBD and
60 ± 15% Acrylonitrile. These results are in good agreement
with our DSC results.

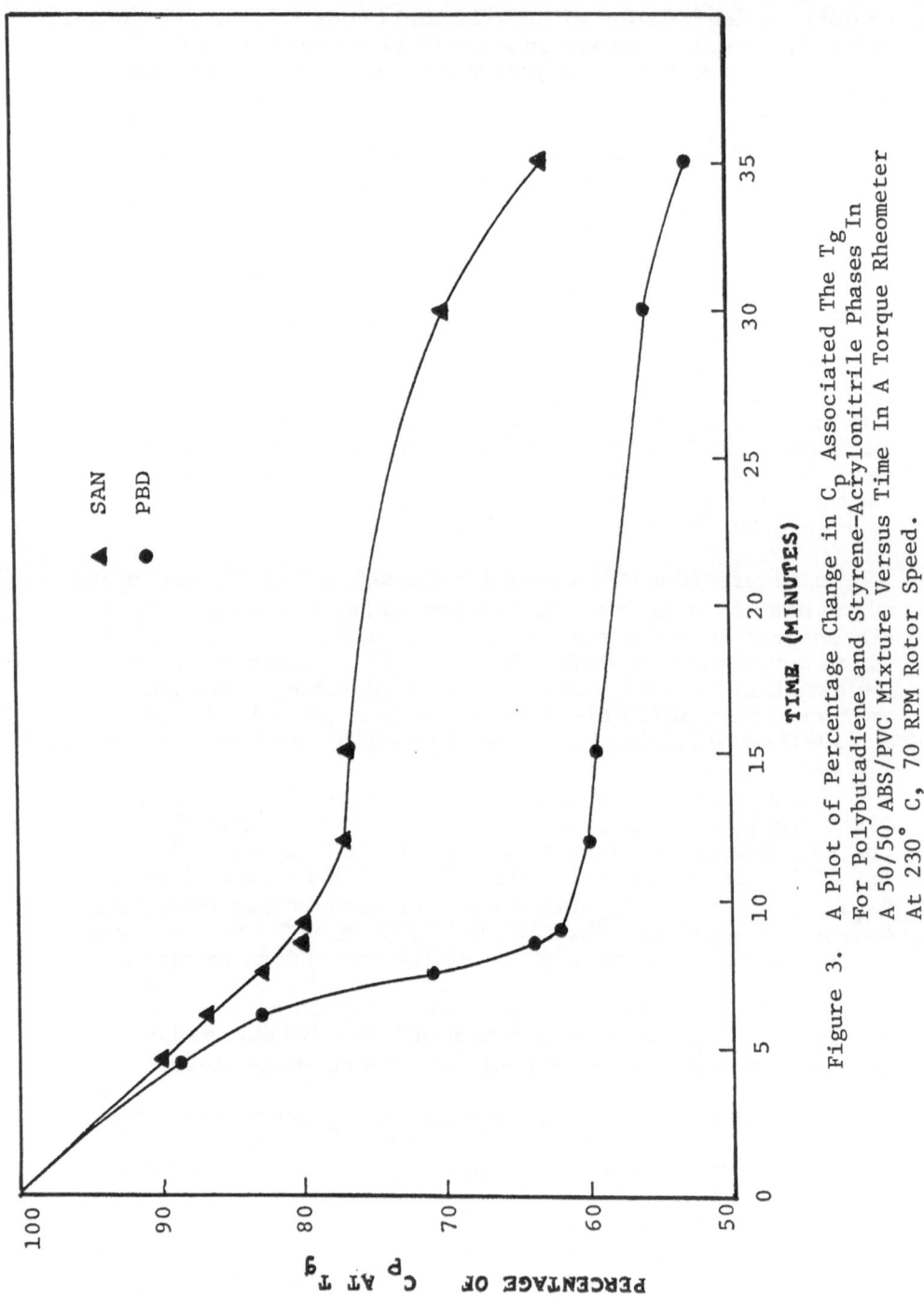

Figure 3. A Plot of Percentage Change in $C_p$ Associated The $T_g$ For Polybutadiene and Styrene-Acrylonitrile Phases In A 50/50 ABS/PVC Mixture Versus Time In A Torque Rheometer At 230° C, 70 RPM Rotor Speed.

TABLE III

Thermogravimetric Weight Loss Data for PVC

and ABS at 230$^\circ$C in Air

| COMPOUND | 10 Minutes | 35 Minutes |
|---|---|---|
| Cycolac T (1000) | 0.082% wt loss | 1.75% wt loss |
| PVC Compound | 1.04% wt loss | 5.4% wt loss |
| 50% ABS/50% PVC | 1.06% wt loss | 5.3% wt loss |
| Calculated values for 50% ABS/50% PVC | 0.93% wt loss | 3.6% wt loss |

Isothermal gravimetric analysis (230$^\circ$C in air) of the blend compared with the components (See Table 3) showed (after 10 minutes) there was no greater release of volatiles from the blend than would be expected based upon the volatilization behavior of the pure components. However, after 35 minutes, the data indicates a significant increase in volatiles above that predicted for the blend. These results in conjunction with the I. R. data indicates sufficient scission has occurred to yield degradation products volatile at 230°C.

It should be noted that it is the rubber phase which toughens these types of blends and small losses of rubber through process degradation can be expected to yield a material with significally lower impact strength and tensile elongation. [6,7,8] Data from a recent aging study [5] of ABS at 71° illustrates this point. Analysis of the rubber content revealed a change in rubber content from the initial value of 14 weight percent down to approximately 11%. The notched Izod impact strength for this example dropped from above 4 to less than 2 foot lbs per inch of notch.

Typical processing conditions for these blends are normally 30°C or more below the temperatures of this study with residence times of 3 minutes or less so that we would expect the observed effects of thermal degradation to be minimal in normal molding practice. Nevertheless, in a molding shop without automated process control, this type of processing temperature may occur occasionally and could cause product embrittlement.

Conclusions
        During torque rheometer stability studies of blends of PVC and ABS, we have observed a reduction in thermal stability for the blend over each component separately which causes a premature increase in melt viscosity. DSC, TGA and IR Spectroscopic analyses show that this lower stability is a result of thermal degradation in all three (SAN, PBD, and PVC) phases of the blend and most rapidly to the polybutadiene rubber phase.

The present evidence indicates what phases are being degraded but does not indicate how this occurs. In thefuture more work will be done to elucidate both the mechanism and develop a stabilizer package for the blend. In the meantime in order to avoid product embrittlement, due care must be taken to process the ABS/PVC blends at 200°C or less.

## REFERENCES

1.  H. E. Bair, Polymer Engineering and Science, Vol. 10, pg. 247 (1970); H. E. Bair, Analytical Calorimetry, Vol. 2, pg. 51, Edited by T. S. Porter and J. F. Johnson, Plenum Press, N. Y. (1970).
2.  A. P. Weir, D. A. Williams and J. D. Woodcock, Chemistry and Industry, pgs. 990-992, August 28, 1921.
3.  J. B. DeCoste, S.P.E.-J., Vol. 21, pg. 764 (1965).
4.  B. D. Gesner, J. Applied Polymer Science, Vol. 9, pg. 3701 (1965).
5.  H. E. Bair, D. J. Boyle and P. G. Kelleher , To be published.
6.  C. B. Bucknall and D. G. Street, J. Applied Polymer Science, Vol. 12, pg. 1311 (1968).
7.  P. G. Kelleher, D. J. Boyle and B. D. Gesner, J. Applied Polymer Science, Vol. 11, pg. 1731 (1967).
8.  A. Casale, O. Salvatore and G. Pizzigoni, Polymer Engineering and Science, Vol. 15, pg. 286 (1975).

# DYNAMIC VISCOELASTICITY OF WOOD-POLYMER COMPOSITES PREPARED BY AN ELECTRON BEAM ACCELERATOR

T. Handa, S. Yoshizawa, M. Fukuoka and M. Suzuki

Dept. of Applied Chemistry, Faculty of Science
Science University of Tokyo
Kagurazaka, Shinjuku-ku, Tokyo, 162, Japan

## ABSTRACT

The dual characteristics in the performance of polymers in wood systems have been pursued with regard to the resolution of mechanical anisotropy of wood and the improvement in dimensional stability. The increase of the dynamic modulus in the tangential direction of wood ($E_T'$) vs. polymer converged ideally to $E_L'$ in the longitudinal direction of wood at a polymer fraction of 1.0. This increase of $E_T'$ is due to the interaction between polymer and the surface of the cellular parts, as illustrated by the rise of the loss modulus ($E_T''$)-peak ranging from 100°C to 150°C. As for the performance of polymer grafted to the cellular parts, stabilization of wood to moisture was found by noting the decrease of dielectric loss peak at low temperatures (corresponding to the rotational motion of methylol groups of the amorphous cellulose and hemicellulose). As a result, it seems likely that the performance of polymers in wood to strengthen its properties is deeply concerned with the balance between the mechanical gain (increase in $E_T'$ and $E_T''$ of wood) and the improvement in its dimensional stability to reduce the shear by swelling on the kind of polymerization specific to the individual polymers, their molecular structure, their residential sites, and their content in wood.

## INTRODUCTION

A large number of studies have been made so far in this comprehensive area of research. However, our survey has shown that commercially available wood-polymer composite (WPC) products are few. This is due primarily to their cost as compared to other competitive products, and secondarily, due to the difficulty of

263

preventing the formation of small surface cracks caused by fatigue
after long exposure to the environment.  Obviously, the main
objective of producing WPC either by radiation cure, chemical cure,
molding or by any other means is to increase the mechanical properties, inducing decrease in the anisotropy of the dynamic modulus,
E', and increase in the dimensional stability of wood simultaneously.
However,  many problems are encountered in WPC technology.  For
example, repeated cycles of moisture adsorption and desorption
induce more fatigue cracks with a composite of high mechanical
properties, provided that polymer content of the system has been
enhanced regardless of the anti-shrink efficiency (ASE) level.

     It seems likely that the performance of polymers in wood to
strengthen its properties is deeply concerned with the balance
between the increase in the dynamic modulus E' and the loss modulus E" along the tangential direction of wood and the improvement
in dimensional stability.  It is also dependent on the kind of
polymerization, specific type of polymers, the molecular structure,
their location sites in wood, and their content.  Therefore, the
present investigations have been pursued primarily for this purpose
by pursuing the following steps:

> (1)  E"-band isolation in the anisotropic change of
>      the mechanical patterns due to the thermal
>      behavior of wood.

> (2)  The performance of polymers in perturbing the
>      complex higher order structure of wood.

> (3)  The balance between the mechanical gain due to
>      the increase in $E_T'$ and the improvement in
>      dimensional stability to reduce the shear by
>      swelling from water-adsorption.

EXPERIMENTAL

     Veneers used in this study were rotary cut beech ones with
a thickness of 0.65 mm and of the same quality which had been care-
fully selected from factory products.  All samples were dried in
vacuo at 80°C for 30 hrs.  Monomers used in this study were styrene (St), methyl methacrylate (MMA), acrylic acid (AA), and acrylonitrile (AN).  For the WPC (irradiation) system, monomer was
impregnated into woods of various moisture contents.  They were
polymerized by a Van de Graaff type electron beam accelerator with
total dosages of 3, 6, 9 and 12 Mrad, respectively.  After the
irradiation process, the residual monomer in the system was ex-
pelled in vacuo.  For the WPC (injection) system, benzene solutions
of polymer at various concentrations were introduced into wood by
the repeated cycle of freezing, degassing, and melting, and then
dried at 80°C for 30 hrs. in vacuo to expel the benzene completely.

The dynamic modulus E' and the loss modulus E" of specimens
have been measured by a Rheovibron DDV-II and a vibrating reed
method.   The dielectric constant $\varepsilon$ ' and the dielectric loss factor
$\varepsilon$ " of specimens were measured by the transformer bridge-type at
each measuring frequency from 110 Hz to 1 MHz.


RESULTS AND DISCUSSION


Thermal Behavior of Wood

Figure 1 shows the temperature dependence of the E' and E"
at 110 Hz as a function of grain angle between tangential (T)-
direction and longitudinal (L)-direction to the fiber axis of wood.
Corresponding to the decrease in E', the specific E"-peaks are
recognized in the temperature range from 230°C to 260°C vs. the
grain angle, respectively.

The thermal expansion of the wood vs. temperature along the
L-direction and T-direction is illustrated in Figure 2.   Above
230°C, shrinkage in the T-direction and the resulting high elongation in the L-direction are observed.   The turning temperature at
ca. 230°C of thermal elongation and shrinkage of the sample in the
T-direction can be assigned to the increase of the labile portion
by the oxidative degradation of cellulose chains in the amorphous
portion of the higher order fibril structure in conjunction with
thermal breaking of the intermolecular hydrogen bond which makes
the fibril angle to the L-axis maximum.

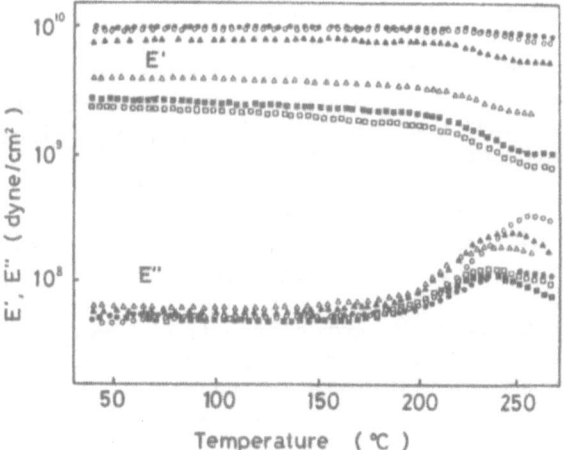

Figure 1.   Temperature dependence of the dynamic modulus E' and
            the loss modulus E" at 100 Hz as a function of the grain
            angle of wood.   Grain angle: ●, 0°(L); O , 11.3°; ▲ ,
            22.5°; Δ , 45°; ▆ , 67.5°; ▢ , 90°(T).

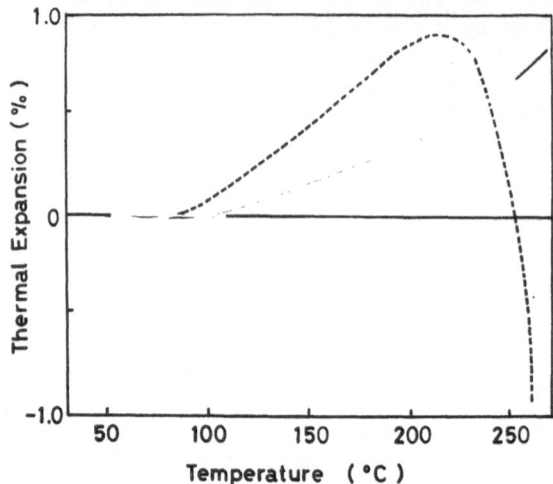

Figure 2.   Temperature dependence of the thermal expansion for L-
            and T-direction of wood during the measurement of the
            dynamic viscoelasticity. ——— , L-direction ----- ,
            T-direction.


     Figure 3 shows that the (002) spacing in the cellulose
crystallite of wood increases linearly with temperature, and the
variation of the half width angle of (002) representing the cellulose fibril orientation is also noted from the X-ray diffraction
patterns of wood (1). The strong increase in the spacing of (002)
and also the decrease in the half width of (002) vs. temperature
above 230°C correspond to the thermal shrinkage in the T-direction
and the compensated strong elongation in the L-direction, which
represent the reduction of the fibril angle regarding the fibril
orientation in the cell wall.

     Finally, on the basis of the results so far, four E" bands
in the high temperature region are selected and assigned as common
and intrinsically peculiar to wood (Figure 4).

     Band (a) at ca. 255°C comes from the decrystallizing
     structural change to stretch fibrils in a way so as
     to reduce the fibril angle by the dislocation due to
     the thermal breaks of the crystallite accompanied
     by shrinkage in the T-direction.

     Band (b) at ca. 230°C comes from the degradative
     structural change reflected in the thickening of
     fibrils in a way so that the fibril angle is at

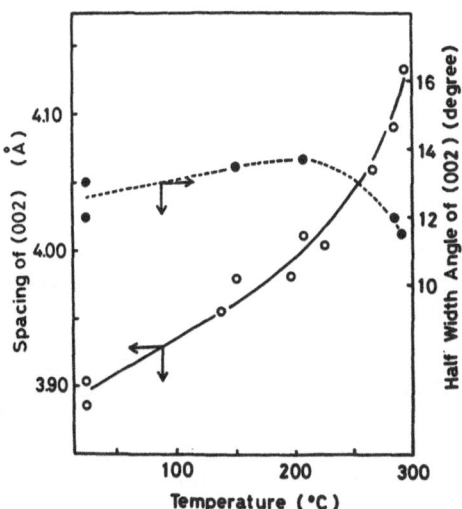

Figure 3.   Temperature dependence of spacing of (002) in cellulose
crystallite and the half value angle of (002) representing cellulose fibril orientation. O , spacing of
(002); ● , the half value angle of (002).

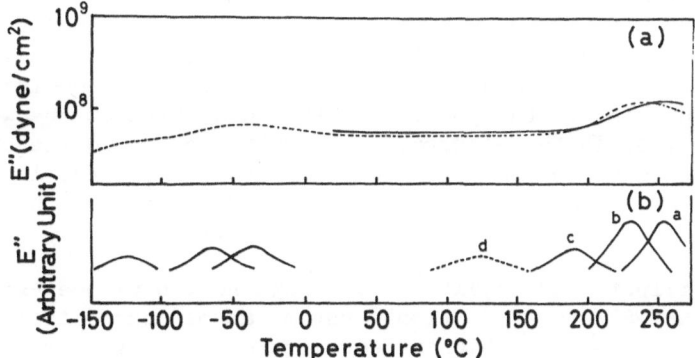

Figure 4.   (a) Temperature dependence of the loss modulus E" at
110 Hz for L- and T-direction of wood. ——— , L-direction;
- - - - , T-direction.   (b) Schematic diagram characterizing
E"-band separated from the temperature dependence of E".

its maximum due to the thermal relaxation of the
intermolecular hydrogen bonds.

Band (c) at ca. 190°C comes from the transition in
the motion of amorphous cellulose and hemicellulose.

Band (d) at ca. 125°C comes from the deformation in
the fibril orientation by the disorder in its conformation close to the surface of the cellular parts.

The apparent activation energy due to band (a), band (b),
and band (c) have been estimated to be 440 kcal/mol, 160 kcal/mol,
and 80 kcal/mol, respectively.


## Performance of Polymers in WPC

The performance of polymers in increasing the E' of wood and
in perturbing the complex higher order structure of wood has been
pursued by taking into consideration the situation of the interaction of polymers with wood as it appeared in the change of the
dynamic viscoelastic and the dielectric patterns of WPC systems.

Figure 5 shows that the E' and E" of the MMA-PMMA-WPC
(irradiation) system increased logarithmically as the polymer
fraction (P.F.) increased.  Test specimens were prepared from the
oven dried wood samples which were cut preliminarily in various
grain angles.  A similar increase in E' and E" as the polymer
fraction increased was recognized in the St-PSt-WPC (irradiation)
system, polyester-WPC (irradiation) system, and solid paraffin-
WPC (injection) system.  The logarithmic increase in E' vs. P.F.
converged finally to a definite value at P.F. of 1.0, while the
increase in E" converged to P.F. of ca. 0.7, corresponding to the
ultimate packing of the polymer in the void parts of the wood.
As the polymer in these systems is hardly present in the cellular
parts because of no elongation of wood by polymer, the increase
of E' and E" is believed to be caused by the interaction between
polymer and the surface of the cellular parts.

As illustrated in Figure 6, regarding the viscoelastic pattern of the PSt-WPC (injection) system, an obvious difference is
recognized between the higher temperature features of the pattern
of WPC and that of wood.  The rise of a new $E_T$"-peak at ca. 125°C
by the injection of polymers has been assigned to the interfacial
structural change of the surface in the cellular parts due to
interaction between polymer and the surface of the cellular parts.
The apparent activation of the thermal motion of polymers trapped
on the surface of the cell wall of wood has been estimated to be
ca. 110 kcal/mol, i.e., approximately 1.5 times higher than that
of the polymers (2).  Injected polymers in wood cannot take part

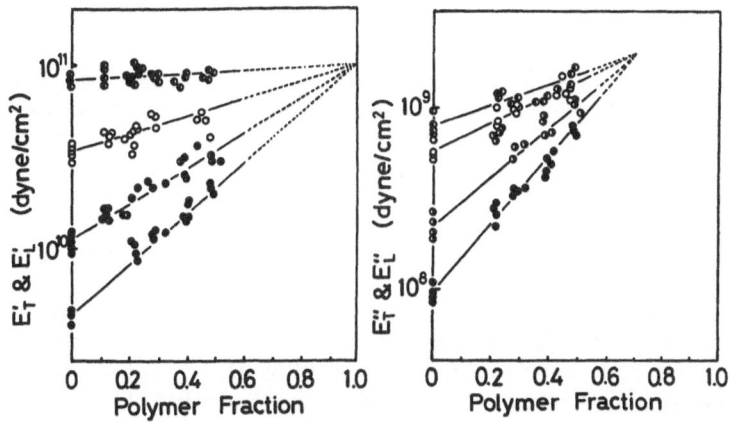

Figure 5.   The change of log E' and log E" for MMA-PMMA-WPC
            (irradiation) system vs. polymer fraction.
            ◑ , 0°(L);   ○ , 22.5°;   ◐ , 45°;   ● , 90°(T).

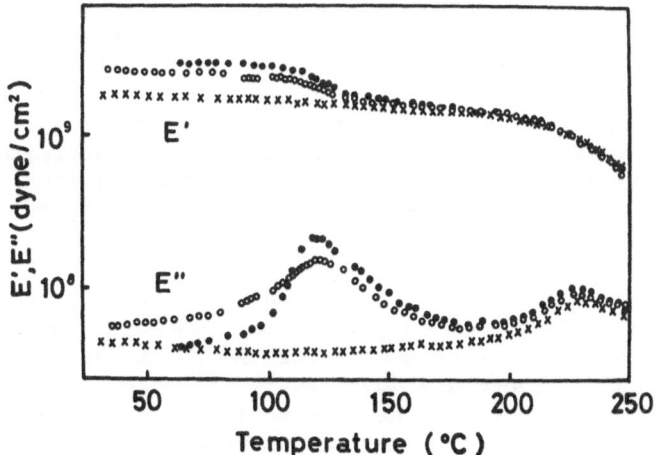

Figure 6.   Temperature dependence of the dynamic modulus $E_T$' and
            the loss modulus $E_T$" at 110 Hz for PSt-WPC (injection)
            system with polymer content of ca. 25%. ○ , WPC;
            ● , heat treated WPC;  X , wood.

in the structural change of wood as concerns the $E_T$"-peak at ca.
230°C because of the difficulty in their geometrical location.

Regarding the performance of grafted polymer induced in wood
under electron beam irradiation, Figure 7 shows that the lower
temperature of $\mathcal{E}$ " peaks, based on the rotational motion of
methylol groups in the amorphous cellulose and hemicellulose (3-5),
are observed to be shifted in the PSt-WPC (irradiation system compared to wood.

As illustrated in Figure 8, a specific lowering of the
apparent activation energy for the rotational motion of methylol
groups is observed for the PSt- and PMMA-WPC (irradiation) systems
vs. polymer content. This means that the more the polymer is
grafted to the system (6), the more active the rotational motion
of methylol groups will be. HOwever, the feasibility of the
rotational motion of methylol groups depends on the specificity
of the bulkiness of the system vs. the kind of polymer. It is
also reflected in the difference of the situation of Cole-Cole's
plots between wood and WPC (irradiation, injection) systems (7,8).
Grafted polymers make the cellular parts random by internal shear
due to bulkiness and helps the rotational motion of methylol
groups, while injected polymers are not related to the rotational
motion of methylol groups.

PSt and PMMA formed by electron beam irradiation have no
effect on the crystallinity of cellulose insofar as the blackness
in the X-ray diffraction patterns of both WPC (irradiation) systems
taken at room temperature are concerned. However, the high temperature feature of the $E_T$" pattern of the PSt-WPC (irradiation)
system indicates the enhancement of band (c) at ca. 190°C and the
lowering of bands (a) and (b) (Figure 9). Therefore, it seems
likely that grafted PSt somewhat perturbs the crystallites and
increases the amorphous portion in wood by thermal shear at high
temperatures.

Figure 10 shows an obvious difference between PSt and PMMA
in the mode of the lowering in temperature of the $E_T$" peak at ca.
230°C by the grafted polymer. This corresponds to the lowering
of the apparent activation energy for the rotational motion of
methylol groups. These are governed by the specificity based on
their molecular volume and the orientation of their unit to cellulose chains regarding both grafted polymers.

Figure 11 shows the viscoelastic patterns of the PAA-WPC
(irradiation) system. Bands (a) and (b) are completely perturbed.
This means that a considerable portion of crystallites in fibrils
must be destroyed due to the thermal shear based on the specificity
of this polar polymer. Therefore, the complete collapse of the

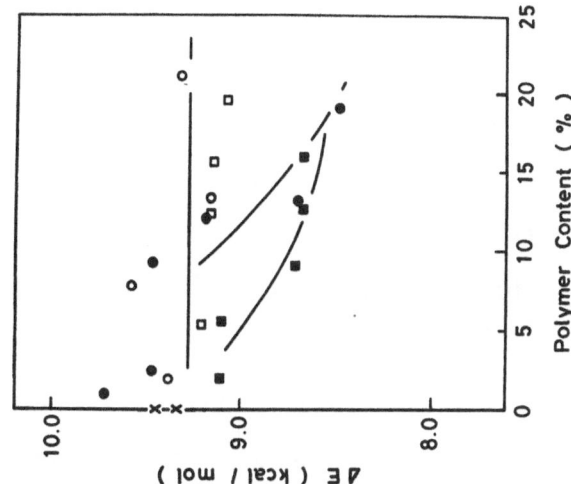

Figure 8.    The apparent activation energy corresponding to the rotational motion of methylol groups ($\Delta$ E) vs. polymer content.  O, PMMA-WPC (injection) system; ●, PMMA-WPC (irradiation) system; □, PSt-WPC (injection) system; ■, PSt-WPC (irradiation) system; ✗, wood.

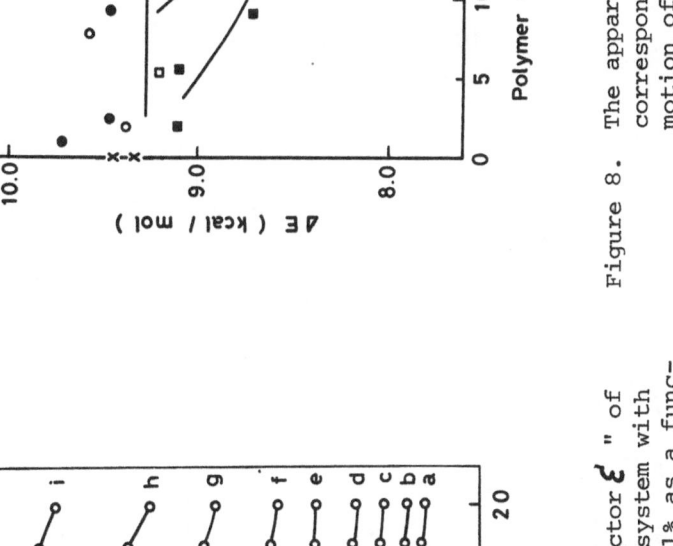

Figure 7.    The dielectric loss factor $\varepsilon$ " of PSt-WPC (irradiation) system with polymer content of 16.1% as a function of temperature at respective frequencies.  Measuring frequency: a, 110 Hz; b, 300 Hz; c, 1 kHz; d, 3 kHz; e, 10 kHz; f, 30 kHz; g, 100 kHz; h, 300 kHz; i, 1 MHz.

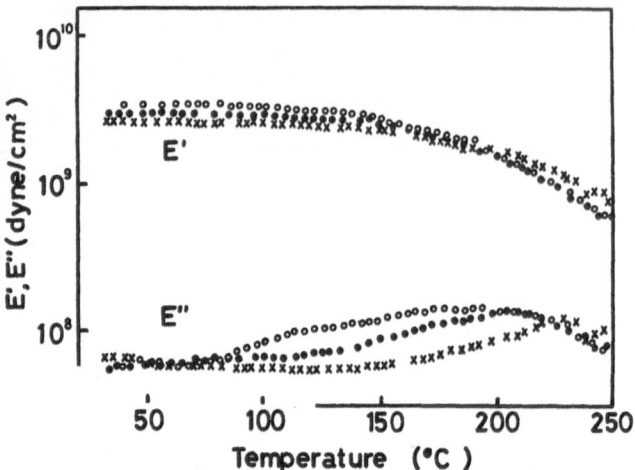

Figure 9.   Temperature dependence of the dynamic modulus $E_T'$ and
the loss modulus $E_T''$ at 110 Hz for PSt-WPD (irradiation.
system with polymer content of ca. 20%. $O$ , WPC; ● ,
homopolymer extracted WPC; $X$ , wood.

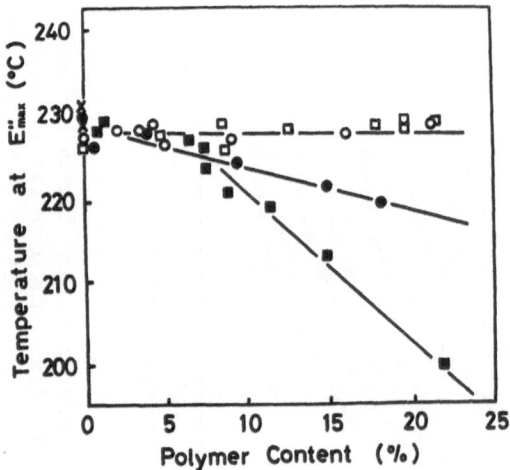

Figure 10.   The shift of the temperature at $E_T''$ max in 200-230°C
vs. polymer content. $O$ , PMMA-WPC (injection) system;
● , PMMA-WPC (irradiation) system; $\square$ , PSt-WPC
(injection) system; ■ , PSt-WPC (irradiation) system;
$X$ , wood.

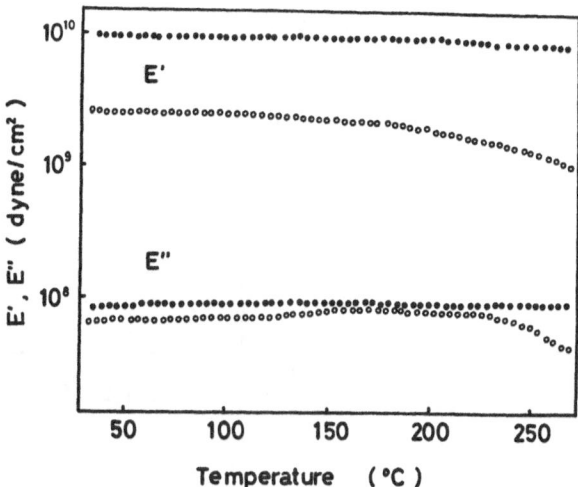

Figure 11.   Temperature dependence of the dynamic modulus E' and
the loss modulus E" at 110 Hz for PAA-WPC (irradiation)
system with polymer content of ca. 16%.  ●  , L-direction;  ○  , T-direction.

peak is considered to be caused by the internal thermal shear imposed on the system.

## Balance Between the Increase in $E_T'$ and the Dimensional Stability

Figure 12 shows that the lowering of $E_T'$ and the compensative
increase in $E_T''$ are exaggerated by the increase in polymer content.
This lowering of $E_T'$ is considered to be due to the presence of
the internal small cracks which are induced by the deformation shear
between the swollen cellular parts and those adherred to polymers.

Figure 13 shows that anti-shrink efficiency (ASE) for the
PAN-WPC (irradiation) system and the PMMA-WPC (irradiation) system
increases up to ca. 70% with increase of polymer.  In order to prevent the formation of cracks by moisture adsorption, an improvement
of ASE of the system by the combination of polymers with cellular
parts of wood is necessary.

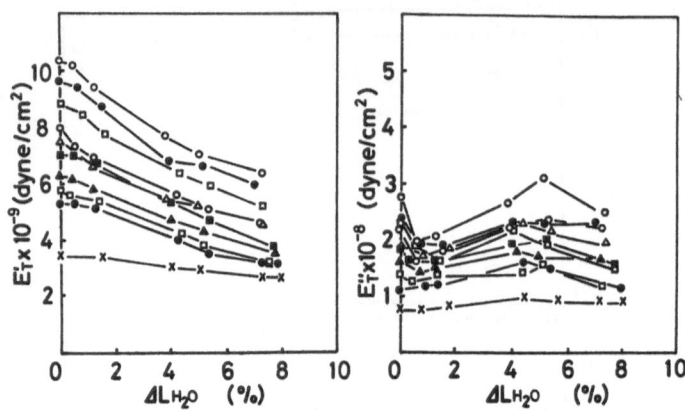

Figure 12.  The change of the dynamic modulus $E_T'$ and the loss
modulus $E_T''$ for PSt-WPC (injection) system vs.
$\Delta L_{H_2O}$ by moisture adsorption.○●□■△▲ WPC;
✗ , wood.

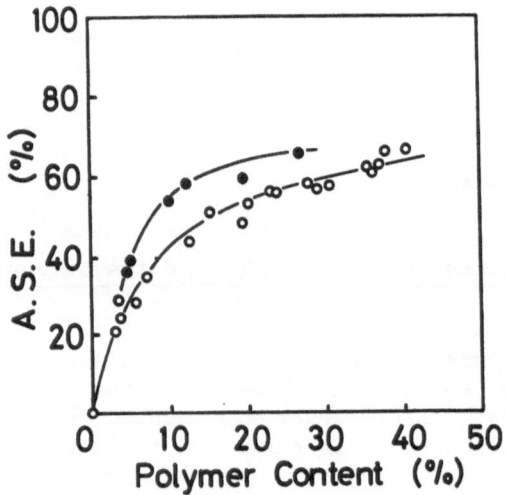

Figure 13.  Effect of polymer content on anti-shrink efficiency
(ASE) for the PAN-WPC (irradiation) system and the
PMMA-WPC (irradiation) system. ● , PAN-WPC (irradi-
ation system; ○ , PMMA-WPC (irradiation system.

    The fatigue cracks of WPC, due to repeated moisture attack,
are considered proportional to the internal work of the strain
induced by moisture.  The estimation of the internal work made by
the swelling strain due to the change in moisture was made in terms
of the hysteresis in $E_T'$ x $\Delta L_{H_2O}$ vs. swelling of the system during
the quasi-static cycle of environmental relative-humidity (which
was controlled by the moisture content of the system concerned).
The internal work due to the internal strain is expected to be proportional to the dynamic modulus in the dried state, Eo', and ASE,
experimentally confirmed by the relation in Figure 14.

    Finally, the following relation was established from the plot
of internal work, Sw vs. $(\Delta L_{H_2O})^2$,

$$Sw = kEo' \text{ x } (\Delta L_{H_2O})^2$$

where Eo' is the dynamic modulus of WPC specimens in the dried
state and k is the proportionality constant which is a function of
the polymer specimens and the molecular weight.

    Therefore, if 100% gain in Eo' is obtained by increasing the
polymer content in the system, the dimensional change due to
moisture adsorption should be suppressed to 1/2 of the original
system by increasing the polymer content in the cellular parts.

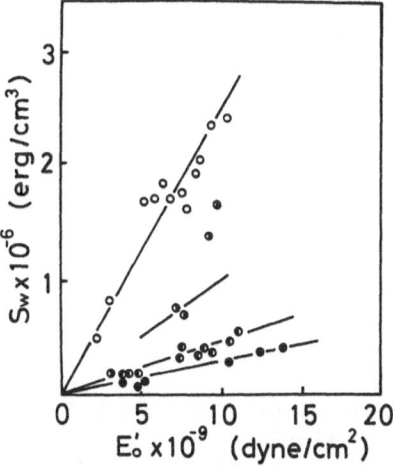

Figure 14.   The change of the induced apparent
             internal work per unit volume, Sw vs.
             the dynamic modulus of PAN-PSt-WPC
             (irradiation-injection) system in
             the dried state, Eo', before the
             cycle of wetting and drying.
             ASE: ○ , 0%; ◐, 37.7%; ◑ , 53.3%;
             ● , 65.0%.

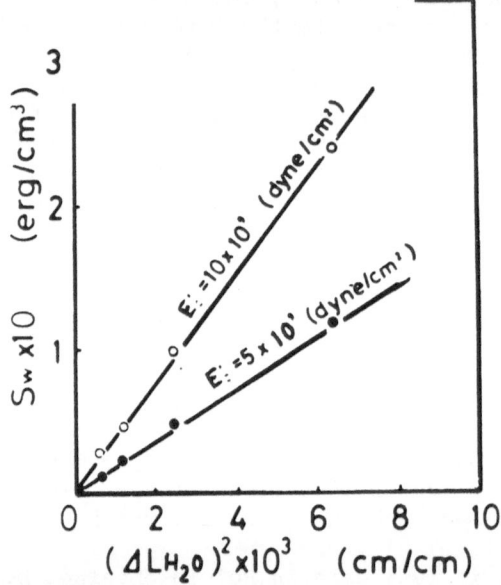

Figure 15.  The change of the induced
apparent internal work per
unit volume, Sw vs.$\Delta$ $L_{H_2O}$
by moisture adsorption
for PAN-PSt-WPC (irradia-
tion-injection) system.

REFERENCES

1.  T. Handa, S. Yoshizawa, M. Suzuki and T. Kanamoto, Kobunshi
    Ronbunshu, 35, 117 (1978).
2.  L. E. Nielsen, "Mechanical Properties of Polymers," Reinhold
    Pub. Corp. (1962).
3.  G. P. Mikhailov, A. I. Artyukhov and T. I. Borisoba, Polym.
    Sci. USSR, 9, 2713 (1967).
4.  G. P. Mikhailov, A. I. Artyukhov and V. A. Shevelev, Polym.
    Sci. USSR, 11, 628 (1969).
5.  M. Norimoto and. T. Yamada, Wood Research, 52, 31 (1972).
6.  T. Handa, S. Yoshizawa and K. Hatakeyama, J. Appl. Polym.
    Sci., 23, 1527 (1979).
7.  T. Handa, S. Yoshizawa and M. Fukuoka, Kobunshi Ronbunshu,
    34, 617 (1977).
8.  T. Handa, S. Yoshizawa and M. Fukuoka, Kobunshi Ronbunshu,
    35, 307 (1978).

# INDEX